# ELECTRONIC ENERGY BANDS IN SOLIDS

L. PINCHERLE
Professor of Mathematical Physics
Bedford College, University of London



MACDONALD : LONDON

SBN 356 03176 4

First Published in 1971 by
Macdonald & Co. (Publishers) Ltd
49/50 Poland Street,
London, W.1





PRINTED AND BOUND IN ENGLAND BY
HAZELL WATSON AND VINEY LTD
AYLESBURY, BUCKS

UNIVERSITY PHYSICS SERIES

*Series Editor:* Professor B. Donovan, Westfield College, University of London

# ELECTRONIC ENERGY BANDS IN SOLIDS

# Contents

# Editorial Foreword

The University Physics Series consists of compact and authoritative monographs on well-defined aspects of pure and applied physics, with particular reference to fields in which research is currently active. Covering a wide range of topics, the series, although addressed primarily to physicists, will be of interest also to chemists, mathematicians, metallurgists and engineers. In general, the emphasis is on thoroughness of exposition rather than on strict uniformity in the level of treatment, and concessions to the reader will be largely dependent upon the character of the individual topic. The tendency to broaden the scope of first degree physics courses has been accompanied by a remarkable proliferation of graduate courses, in both the main stream of physics and the application to neighbouring fields. In relation to such graduate courses these monographs will be especially useful and in many cases they will be relevant also to the more specialised areas of undergraduate courses in the physical sciences.

March 1971 B. Donovan

# Preface

This book grew out of lectures given at a M.Sc. course in solid-state physics and thus aims at bridging the gap between the elementary undergraduate introduction to the properties of the electronic energy levels in solids and the specialised accounts for theoretical researchers.

It then gives, within the one-electron approximation, the quantum theory of electrons in perfect lattices (band theory), at a level which should be readily accessible to any graduate in physics or engineering. It tries to keep a balance between a formal exposition of the theory and simple applications, and also to touch on a variety of items more or less connected with band theory.

Some knowledge of the fundamental concepts is assumed, but most of them are re-defined and discussed. It was more of a problem to decide what to assume in the way of mathematical knowledge and no doubt some readers will think that unnecessary mathematical explanations have been included, while others would have wanted more details.

The question of units is not acute in band theory, since atomic units are the obvious choice. When macroscopic quantities are involved, cgs units have been used, in order to conform with most textbooks on solid-state physics.

Thanks are extended to the publishers and authors who have allowed reproduction of diagrams: particular acknowledgement is made in the text in each case. I am also grateful to Professor B. Donovan and Dr R. Mansfield for useful discussions.

March 1971 L. Pincherle

# 1

# Foundations of band theory

## 1.1 One-electron approximation. Hartree-Fock equations

Band theory is the study of the grouping of the electronic energy levels in solids into energy bands and of all the consequences that follow from this fundamental fact. It is based on the one-electron approximation, which assumes that each particle can be described by an individual wavefunction. It is important to the solid-state physicist because its concepts and language are adequate to describe a very large amount of experimental information. The band structure of a material depends on the structure of the crystal, on the species of atoms in the crystal and on the electronic states of the atom or atoms from which the bands derive. Except when otherwise stated, perfect crystals will be considered.

The one-electron theory starts from the Hartree-Fock equations, though the concept of bands in no way depends on this approach. It might be thought that modern many body theory goes far beyond the Hartree-Fock scheme. Indeed it does, and it has shown when and how to use the Hartree-Fock theory, but the result, as far as energy bands are concerned, has been that this theory is a reliable and useful technique. We shall not consider here the modifications and improvements that have been suggested to the original theory (see for instance Slater, 1957).

The basic idea of the Hartree-Fock method is to use a Slater-Fock determinant

$$\Psi(\mathbf{r}_1 \ldots \mathbf{r}_N) = \frac{1}{\sqrt{N!}} \begin{vmatrix} \Psi_1(\mathbf{r}_1) & \ldots & \Psi_1(\mathbf{r}_N) \\ \Psi_2(\mathbf{r}_1) & \ldots & \Psi_2(\mathbf{r}_N) \\ \cdot & & \\ \cdot & & \\ \cdot & & \\ \Psi_N(\mathbf{r}_1) & \ldots & \Psi_N(\mathbf{r}_N) \end{vmatrix} \tag{1.1}$$

where the $\Psi_i$ are normalised one particle wavefunctions and the coordinate

$\mathbf{r}_j$ denotes both space and spin variables of one particle. $N$ is the total number of electrons.

This is the antisymmetric wavefunction which comes closest to obeying the hypothesis of statistical independence of all possible one-electron quantities (probability density, current ...). In the original Hartree method, only the simple product corresponding to the principal diagonal of the determinant was used. If several configurations are possible, a combination of Slater determinants must be taken, the squares of their coefficients being proportional to the probability of the corresponding configuration.

Using (1.1) we then find the eigenstates, generally only the lowest eigenstate, of the Hamiltonian $H$ for the system of interacting electrons. For simplicity the non-relativistic Hamiltonian is considered, with only the electrostatic interaction between the electrons

$$H = \Sigma_j \left[ \frac{p_j^2}{2m_0} + V(\mathbf{r}_j) \right] + \tfrac{1}{2}\Sigma_{j,k} \frac{e^2}{|\mathbf{r}_j - \mathbf{r}_k|} \tag{1.2}$$

where $m_0$ is the free electron mass, and $V$, in a crystal, is the potential due to the atom cores. It is in fact a potential energy, but it is customary to call it simply potential. The resultant equations, derived by standard application of the variational principle, must be solved self-consistently. A set $\Psi_m$ of one-electron functions is assumed, and a new set $\Psi_i$ is calculated by solving wave equations (the Hartree-Fock equations) in which the interaction $\dfrac{e^2}{|\mathbf{r}_j - \mathbf{r}_k|}$ is replaced by its average over the probability distribution of all the other electrons, so that we have an effective one-electron potential. In principle, the procedure is repeated with different initial functions until the final $\Psi_i$ are identical with the initial $\Psi_m$. The equations assure that the resultant total energy is an extremum relative to all neighbouring Slater determinants.

Neglecting spin, the Hartree-Fock equations for the one-electron functions $\Psi_i(\mathbf{r}_1)$ are (for a derivation see for instance Slater, 1957; Anderson, 1963):

$$\left[ -\frac{\hbar^2}{2m_0}\nabla_1^2 + V(\mathbf{r}_1) + \Sigma_j e^2 \int \frac{|\Psi_j(\mathbf{r}_2)|^2}{|\mathbf{r}_1 - \mathbf{r}_2|} d\mathbf{r}_2 \right] \Psi_i(\mathbf{r}_1)$$
$$- \Sigma_j\, e^2 \left[ \int \frac{\Psi_j^*(\mathbf{r}_2)\Psi_i(\mathbf{r}_2)}{|\mathbf{r}_1 - \mathbf{r}_2|} d\mathbf{r}_2 \right] \Psi_j(\mathbf{r}_1) = E_i \Psi_i(\mathbf{r}_1) \tag{1.3}$$

If spin is taken into consideration through the Pauli spin operators $\boldsymbol{\sigma}_i$, the operator $\dfrac{e^2}{|\mathbf{r}_1 - \mathbf{r}_2|}$ in the second integral is replaced by $\dfrac{e^2}{2|\mathbf{r}_1 - \mathbf{r}_2|}[I + \boldsymbol{\sigma}_1 . \boldsymbol{\sigma}_2]$, where $I$ is the unit operator. $\Psi_i$, $\Psi_j$ become column vectors with two components. $\boldsymbol{\sigma}_1$ operates on $\Psi_j$, $\boldsymbol{\sigma}_2$ on $\Psi_i$ and the integrals include a summation over the spin indices.

The two sums express, respectively, the average electrostatic potential energy $V_C$ of all the electrons, and the exchange interaction. The two sums include $j = i$ because the $j = i$ terms are identical and cancel. The energy $E_i$ is the negative of the energy required to remove an electron in state $i$ from the system (the electron system must be very large and the one-electron functions must be non-localised – Koopman's theorem). Of course, the total energy is not just the sum of the one-electron energies $E_i$ because we have interacting particles, but depends on the distribution of the particles. It is generally assumed that the space parts of the wavefunctions for opposite spin are identical. This is not always true, for example in some magnetic problems. When the two functions are taken as neither identical, nor orthogonal, one speaks of the unrestricted Hartree-Fock approximation.

Since electrons repel one another, their motions are not independent: they tend to avoid each other. This is expressed by saying that the Coulomb repulsion correlates the individual electronic motions by reducing the probability that two electrons may approach closely. This correlation produces also a 'correlation energy', defined as the difference between the total energy calculated by taking into proper account the Coulomb correlation, and the Hartree-Fock energy. This correlation energy is particularly important in determining the cohesion energy of metals. A correlation effect is the polarisation of the atom core by the outer electrons. Similarly, the dielectric polarisation around an impurity in a crystal is missing in the Hartree-Fock approximation.

In the simple Hartree method, correlations are completely ignored. The Hartree-Fock method also neglects proper Coulomb correlations, but includes correlations, due to the Pauli exclusion principle, between electrons with parallel spins. In fact, the determinant (1.1) vanishes if electrons 1 and 2 have the same space and spin coordinates (but not if the spin coordinates are different, even if two electrons are in the same place). Each electron may be imagined as surrounded by a 'hole', called the exchange hole, or Fermi hole, in the distribution of electrons with parallel spins. The total charge of the exchange hole is $+e$.

Strictly the Coulomb correlation should be taken into consideration before applying the Pauli principle. This can be done very roughly within the one-electron approximation, and must be worked out by a different approach, such as the theory of plasma oscillations (see, for instance, Raimes, 1961). This theory also offers the simplest justification of the independent particle approximation.

One might think that the one-electron energies could be used to approximate the excitation energies of the crystal. In a semiconductor relatively few electrons are excited and they are far apart, hence it seems plausible to treat them as independent and to consider that the excitation energy of the crystal is the sum of the excitation energies of the excited electrons. However, the one-electron energies from the equations (1.3) cannot be used because of the fact that the charge of the exchange hole is one for an occupied orbital, but zero for an excited one. This difference lowers the occupied levels with respect to the excited levels. Occupied and unoccupied levels are pushed apart so that in a metal deficiency of levels results near the Fermi surface. In fact, in the Hartree-Fock scheme the density of states at the Fermi level is zero (see for instance Raimes, 1961). This is contrary to experiment, so that the naive assumption that the excitation energy of the crystal can be expressed as the sum of one-electron energies cannot be maintained. It is found that one should ignore exchange, or at least the dependence of the exchange energy on momentum. This led Slater to suggest that the Hartree-Fock exchange term may be approximated by a weighted average. But it is impossible to give an *a-priori* rule as to how to set up a potential such that the solutions of Schrödinger's equation will give the best one-electron energies to be used in conjunction with Fermi statistics.

The Hartree-Fock equations are a complicated system of $N$, or $2N$, integro-differential equations. The exchange term is the most awkward as it couples to $\Psi_i$ all the other $\Psi$s. A simplification often adopted was suggested by Slater (1951): an 'exchange potential energy' is introduced

$$V_{\mathbf{exch}}(\mathbf{r}_1) = \frac{e^2}{\Psi_i(\mathbf{r}_1)} \Sigma_j \, \Psi_j(\mathbf{r}_1) \int \frac{\Psi_i(\mathbf{r}_2)\Psi_j(\mathbf{r}_2)\, d\mathbf{r}_2}{|\mathbf{r}_1 - \mathbf{r}_2|} \tag{1.4}$$

often evaluated by some drastic approximation, such as the assumption of free electrons. However, the energy levels found by a band calculation based on a simplified exchange potential do not necessarily obey Koopman's theorem and the problem of the 'Koopman corrections' has not yet been solved satisfactorily. The total charge corresponding to the exchange potential is $+e$, the charge of the Fermi hole.

The exchange term now becomes $-V_{\mathbf{exch}}(\mathbf{r}_1)\Psi_i(\mathbf{r}_1)$ and we are left with a separate Schrödinger equation for each $\Psi_i(\mathbf{r}_1)$:

$$\left[-\frac{\hbar^2}{2m_0}\nabla^2 + V + V_C - V_{\mathbf{exch}}\right]\Psi = E\Psi \tag{1.5}$$

All potentials have the periodicity of the lattice in a perfect crystal. The potentials $V_C$ and $V_{\mathbf{exch}}$ as derived from the Hartree-Fock equations depend on the state $i$ considered, so that an index $i$ should be added, but, in order to discuss more simply the general principles of band theory, we shall assume that they are the same for all states. The dependence on $i$ would produce the complication that the one-electron functions would not be orthogonal. The question of the construction of the potential is discussed briefly in Section 5.7.1. We shall from now on indicate simply by $V$ the sum $V + V_C - V_{\mathbf{exch}}$.

If spin-orbit coupling is to be considered, one adds to the operator on the l.h.s. of (1.5) the term

$$\frac{e^2}{4m_0^2c^2}\,\mathbf{p}.\boldsymbol{\sigma} \times \nabla V \tag{1.6}$$

where $\mathbf{p}$ is the momentum operator and $\boldsymbol{\sigma}$ the Pauli spin vector. This term is generally treated by perturbation theory.

The potential of, say, a conduction electron in a metal, along a line joining identical atoms, has the form illustrated in Fig. 1.1. It is essentially because of the quantum-mechanical tunnel effect that electrons in an energy level as

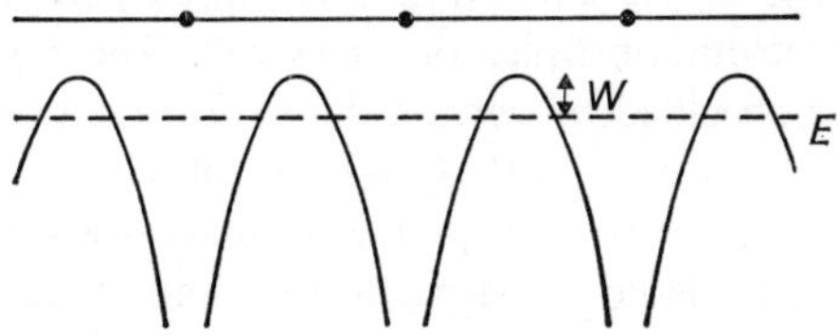


Fig. 1.1. Periodic potential along a line of equal atoms. In metals the potential is approximately constant in the regions between the ion cores.

the one indicated can move through the crystal, but the transparency of the interatomic potential barrier decreases roughly exponentially with the depth $W$ of the level under the top of the barrier, so that electrons bound by more than a few eV can be considered to remain attached to the original atom, and their effect is only that of screening the nuclear charge acting on mobile electrons.

Note that perfect periodicity is of course prevented by the existence of the surface of the crystal, where there is a potential barrier opposing the escape of electrons, but the surface is irrelevant in calculating the electronic energy levels in the bulk, and in fact its existence is replaced by more convenient boundary conditions, such as periodic boundary conditions (Section 2.2), which are in fact equivalent to assuming a crystal of infinite extension.

More important are the deviations from periodicity due to lattice defects, such as vacancies, or interstitial atoms, or dislocations; to the presence of substitutional impurities; to the vibrations of the lattice. In an introduction to band theory, these deviations are not considered. They cause the electrical resistance of the crystal, which, if absolutely perfect, would not impede at all the motion of electrons. Theoretically, they are generally treated by perturbation methods.

The single electron approximation reflects the experimentally well-established fact that electrons and holes in crystals behave in most common cases very much like free electrons and free positrons. As we shall see, the generalisation of the concept of free carriers is conceptually very simple, even if sometimes the effect of the crystal field is very large. It is only necessary to replace the free electron mass by an effective mass $m^*$ (or a tensor generalisation) and to write the Coulomb interaction between two charges in the crystal as $E_C = q_1 q_2/\varepsilon r$, where $\varepsilon$ is the static dielectric constant. In the case of semiconductors the use of $m^*$ is justified because (see for instance Kohn 1957), if we have a system of $N$ electrons occupying a full band, plus or minus one carrier, one of the quantum numbers, $p$, plays the role of a momentum; and, if $E_0$ is the energy of the $N$ electrons, the energy of the system of $(N + 1)$ electrons is $E = E_0 + p^2/2m^*$; thus $m^*$ is really a property of the entire electron system. The single excitation is a 'quasi-particle' which may be pictured as an electron, or hole, of mass $m^*$. These quasi-particles are described by bands specified by certain dispersion relations between energy and momentum. The many-body system is of course capable of other excitations, such as collective oscillations (plasma oscillations) of much higher energy; but the single quasi-particle excitations are the only ones of interest in band theory.

The situation is similar to renormalisation in quantum electrodynamics. The replacement of $m$ by $m^*$ is a mass renormalisation, while the replacement of a charge $q$ by $q/\sqrt{\varepsilon}$ is the analogue of charge renormalisation. In quantum electrodynamics, an electron is really a very complicated excitation of the electron and photon fields, but for many purposes it can be completely described by the experimentally observed mass and charge. Similarly an electron or hole in a semiconductor constitutes a very complicated disturbance, yet in most situations it can be completely characterised by an effective mass and charge. The essential point is that the energy bands defined starting from the many-body Schrödinger equation are the same as those obtained from the Hartree-Fock equations in the one-particle model.

Even in metals, in which interactions are more important than in semiconductors, the conclusions of the one-particle approximation are applicable to a surprising extent. In particular, a well defined Fermi surface, enclosing the region of momentum space occupied by nearly free electrons described by one-particle wavefunctions, exists, and its detailed characteristics can be determined with great precision. We shall examine in Section 5.4 an argument based on 'pseudopotentials' which tries to explain this remarkable state of affairs.

## 1.2 Band approach and valence bond approach

In the band approach the electrons are assumed to move through the whole crystal, and any electron has the same probability of being at equivalent points in the crystal. In the valence bond theory, on the other hand, electrons are considered to be localised, each being bound to a certain nucleus, or part of a chemical bond. This last theory is the extension to crystals of the Heitler-London approach for molecules, in which the molecule is thought to arise from the coming together of atoms complete with all electrons, atoms which then interact. However, in the molecular orbitals approach, of which the band model is the extension, only the nuclei and inner electron shell are first thought to come together, and then the valence electrons are thought to move in polycentric orbits extending through the whole molecule.

As an illustration, consider the molecule $H_2$. Both methods start with the Slater determinant

$$\Psi = u(\mathbf{r}_1)v(\mathbf{r}_2) \pm v(\mathbf{r}_1)u(\mathbf{r}_2) \tag{1.7}$$

where spin has been neglected, so that both the symmetric and antisymmetric combinations of products of individual space wavefunctions occur. The two approaches differ in the choice of the functions $u$, $v$. The valence bond theory takes functions centred on the two nuclei, for example, in the

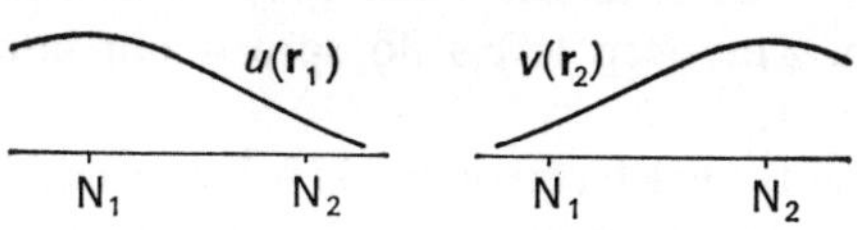


Fig. 1.2. Hydrogen molecule. Electronic wavefunctions in the valence bond approach. (Ground state.)

ground state, functions of the type illustrated in Fig. 1.2 and uses such functions in a perturbation, or variational calculation. The molecular orbitals theory takes $H_2^+$ wavefunctions, for example, in the ground state,

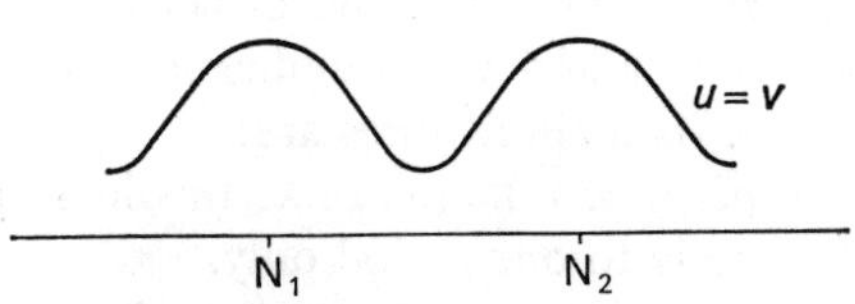


Fig. 1.3. Hydrogen molecule. Electronic wavefunction for the ground state in the molecular orbitals approach.

for both $u$ and $v$ a function like that illustrated in Fig. 1.3, or one antisymmetric about the middle point. Again, this function is used as basis for variational or perturbation calculations. As is well known, the latter treatment gives equal weight to the homopolar and ionic types of bonding, while the valence bond theory in its simplest form considers the binding as purely homopolar.

We shall deal almost exclusively with the band (molecular orbitals) picture, which is most appropriate for simple metals and for the most important semiconductors. Electrons of deep shells, and sometimes *d*-electrons in transition metals, are however better treated as localised, and the valence bond theory is also often more useful for the study of types of bonding, and in describing conduction through a 'hopping' motion. The localised functions that should in principle be used to describe fairly tightly bound electrons are called Wannier functions (Section 3.3).

## 1.3 Origin of bands. Main features of the band approximation

When atoms come together to form a solid, atomic levels must broaden on the assumption that the motion of each electron extends to the whole solid. This can be understood on the basis of the uncertainty principle, since the electrons spends only a short time around each atom. For the outermost levels the spread of the energy must be of the order of a few eV. This does not apply to electrons in deep levels which may be considered as remaining bound to an atom. The deep levels do spread out into bands, but these bands are very narrow.

In a perfect crystal the existence of bands is better understood by adopting the wave picture of the electron, since bands of allowed and forbidden frequencies exist in the case of any wave propagation in a periodic structure (Brillouin, 1946). For electrons, the frequency $\nu$ of the de Broglie waves is not an observable quantity, and the energy $E = h\nu$ is considered instead. In all such cases, for example, in the case of the lattice vibrations, or for electromagnetic waves in a corrugated waveguide, it is found that, considering for simplicity a one-dimensional problem, the dispersion relation, that is the relation between frequency $\nu$ and wavenumber $k$, is of the type indicated schematically in Fig. 1.4. Its main features are:

1. $\nu$ (or $E = h\nu$) is a periodic function of $k$; this indicates that it must be possible to restrict the study to one period only.
2. For any given $k$ there is generally a number of possible frequencies.
3. Values of $\nu$ within certain frequency intervals are not found for any value of $k$.

Yet another way of describing the origin of energy bands is to start with a 'crystal' with two atoms and add successively more atoms. If we have two equal atoms far apart, the levels of the combined system are doubly degenerate. When the atoms approach, the interaction between them splits each atomic level into two molecular levels, removing the degeneracy. This is analogous to the splitting of each frequency when two identical oscillators interact. The argument can be extended to a system of $N$ identical atoms.

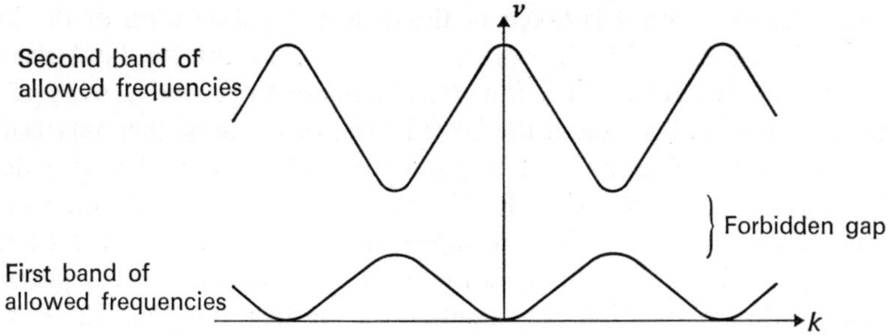


Fig. 1.4. One-dimensional dispersion relation (schematic).

Each addition of a new atom adds one more level and the energies of the $N$ levels are grouped about the original atomic level. In the limit, for $N$ very large, one gets bands, each containing $N$ levels, each band deriving in a sense from an atomic level, though this would often be an oversimplification. This approach shows that energy bands are not restricted to periodic structures, but must exist in amorphous substances, liquids, etc., provided there is overlap between the wavefunctions of electrons on different atoms. The $N$ levels are so close together than in practice they are considered as a continuum and a density of states, $dn_s/dE$ is considered, being the number of levels per unit volume of the crystal per unit energy interval.

To recapitulate, the main approximations of band theory are:

1. Electrons are assigned individual wavefunctions.
2. The action upon any electron of all the other electrons can be replaced by a periodic potential; all deviations from periodicity are treated later by perturbation theory.
3. Anything in the nature of multiplet structure is neglected treating the problem like, in the case of atoms, that of one electron outside closed shells. In this way, all phenomena involving interactions between particles are excluded; for instance, the existence of 'excitons' resulting from the interaction between an electron and a hole.
4. The energy bands are calculated assuming fixed nuclei (adiabatic approximation). The interaction with the lattice vibrations is again introduced at a later stage as a perturbation, and in many cases, particularly at low temperature, is found simply to increase the effective mass of the carriers. In semiconductors, this perturbation, which increases with increasing temper-

ature, produces a decrease of the energy gap $E_G$ with increasing $T$. (The total variation of $E_G$ with $T$ depends also on the variation with $T$ of the lattice parameters.) But when the interaction is very strong, as in ionic crystals, the perturbation treatment is not justified. The appropriate formalism is then that of polaron theory (see for instance Pekar, 1963).

In this theory account is taken of the dielectric polarisation of the ionic crystal by the electric field of the drifting electrons. It is this local polarisation due to the displacement of the ions that increases the effective mass of the carriers. Moreover, because of the large inertia of the ions, this polarisation cannot follow the electron, but forms a local potential well whose depth (0·5–1 eV) is generally sufficient for discrete energy levels of the electron to exist in it. The electrons can be trapped in these states. Thus the trapped electron can maintain with its field the local polarisation, and the potential well due to the polarisation can maintain the electron trapped. These states with the electron localised in a polarisation potential well are called polarons. The potential well, with the electron oscillating in it, can move through the crystal, and it may be proved that this excitation has a dispersion law such that the energy is proportional to the square of momentum. Thus the polaron is another quasi-particle, with the same charge as the electron but an effective mass which can be very different from the free electron mass.

# 2

# The one-electron wavefunctions in crystals

## 2.1 Periodicity conditions

Following the preceding considerations, we want to study the energy levels and wavefunctions of a single electron in an infinite periodic rigid lattice. We thus regard an assembly of many electrons as a collection of non-interacting particles occupying the one-particle levels in accordance with the rules of Fermi statistics.

We have to solve the Schrödinger equation

$$\nabla^2\Psi(\mathbf{r}) + \frac{2m_0}{\hbar^2}\Big[E - V(\mathbf{r})\Big]\Psi(\mathbf{r}) = 0 \tag{2.1}$$

where the potential $V$, including the average field of all the electrons except the one under investigation, is periodic with the periodicity of the lattice. At this stage we neglect spin.

In band theory, units are generally used in which $\hbar = 1$ and the unit of length is the radius of the first orbit of the hydrogen atom,

$$a_0 = \frac{\hbar^2}{m_0 e^2} = 5{\cdot}2915 \times 10^{-9}\ \text{cm}$$

The product $m_0 e^2$ must then be equal to unity. Now, a convenient unit of energy is the Rydberg, that is the ionisation energy of the hydrogen atom $W_H = 13{\cdot}598$ eV $= 2{\cdot}1785 \times 10^{-11}$ erg. Since $W_H = e^2/2a_0$, we take $e^2 = 2$, and then $m_0 = \frac{1}{2}$. These units are useful because all constants are eliminated from Schrödinger's equation which becomes simply

$$(-\nabla^2 + V)\Psi = E\Psi \tag{2.2}$$

The relation between kinetic energy and momentum is $T = p^2$. Since $\hbar c/e^2 = 137{\cdot}039$, the velocity of light in these units is $274{\cdot}078$.

As already remarked, we do not have a boundary conditions problem, but

(2.1) must be solved subject to periodicity conditions, which we wish now to determine.

To gain most simply an insight in the nature of the problem, consider first a one-dimensional problem, that of a linear chain of atoms, of period $a$. Following Born and von Kármán (see, for instance, Born and Huang, 1954), imagine the chain bent round in a large ring of length $L$ containing $N$ periods, so that

$$\Psi(x + Na) = \Psi(x) \tag{2.3}$$

(periodic boundary conditions). Now one of the fundamental assumptions of band theory is that the electron has the same probability of being at equivalent points in the crystal, that is, in this case, at points separated by $ma$, where $m$ is an integer. Thus

$$|\Psi(x + ma)|^2 = |\Psi(x)|^2 \tag{2.4}$$

This means

$$\Psi(x + a) = C\Psi(x), \qquad \Psi(x + ma) = C^m\Psi(x)$$

where $C$ is a complex number of modulus one. When $m = N$, we have

$$\Psi(x + Na) = C^N\Psi(x) \tag{2.5}$$

so that, by comparing (2.3) and (2.5) we find $C^N = 1$, or

$$C = \exp(2\pi im/N) \qquad m = 0, 1, 2, \ldots N \tag{2.6}$$

Thus $\Psi$ satisfies the condition (Bloch, 1928)

$$\Psi(x + a) = e^{ika}\,\Psi(x), \qquad \text{with } k = \frac{2\pi m}{Na} \tag{2.7}$$

Clearly $d\Psi/dx$ satisfies the same condition.
When $N$ becomes very large, $k$ becomes practically a continuous variable. There are $N$ non-equivalent values of $k$, one for every length $2\pi/a$ of what we shall call $k$-space. However, we can add to any $k$ any multiple of $\pm 2\pi/a$ and the relation (2.7) remains unaltered. $k$ is a quantum number, with the dimensions of a wave number, denoting the various states, whose energy also depends on $k$. We shall see that in the simplest cases the totality of states with non-equivalent values of $k$ constitutes an energy band and the different bands are characterised by the particular multiple of $2\pi/a$ (or its three-dimensional generalisation) which has been added to the smallest $k$.

Since electrons can propagate in both directions and we shall see that $\hbar k$ has many properties of momentum, the smallest $k$ should in fact be included not between 0 and $2\pi/a$, but between $-\pi/a$ and $\pi/a$. If all the states of a band are occupied, for each electron with a certain value of $k$ there is another labelled by $-k$. As a consequence, a completely full band carries no electric current.

In three dimensions the argument is similar, by assuming that the crystal is periodic along three fundamental directions, with periods **a**, **b**, **c**, and periodic boundary conditions hold for a parallelepiped of sides $G_1\mathbf{a}$, $G_2\mathbf{b}$, $G_3\mathbf{c}$, where $G_1$, $G_2$, $G_3$ are large integers. It is impossible to bend the crystal to achieve physically these conditions, but they work well as a mathematical device. Note that, if we assumed instead that $\Psi$ vanishes on the surface of the parallelepiped, we would obtain standing, instead of running waves, and these would be of no use in problems of conductivity. Take for simplicity **a, b, c** mutually perpendicular and along the coordinate axes. Then proceeding exactly as in the one-dimensional case, one finds that $\Psi$ must satisfy the condition

$$\Psi_{lmn}(x + a, y + b, z + c) = \exp\,[2\pi i(l/G_1 + m/G_2 + n/G_3)]\,\Psi_{lmn}(x, y, z) \quad (2.8)$$

where $l$, $m$, $n$ are integers. Writing now $k_x = 2\pi l/aG_1$, $k_y = 2\pi m/bG_2$, $k_z = 2\pi n/aG_3$, we have

$$\Psi_{lmn}(x + a, y + b, z + c) = \exp\,(ik_x a + ik_y b + ik_z c)\Psi_{lmn}(x)$$

or for any vector $\mathbf{R}_i = p\mathbf{a} + q\mathbf{b} + r\mathbf{c}$ ($p$, $q$, $r$ integers);

$$\Psi_{\mathbf{k}}(\mathbf{r} + \mathbf{R}_i) = e^{i\mathbf{k}.\mathbf{R}_i}\,\Psi_{\mathbf{k}}(\mathbf{r}) \quad (2.9)$$

Again **k** takes discrete values, but when $G_1, G_2, G_3 \rightarrow \infty$, it may be treated as a continuous variable.

The Bloch relation (2.9) is the fundamental periodicity condition for the wavefunctions in a perfect crystal and Schrödinger's equation must be solved subject to this condition. It states that, in going from **r** to an equivalent point $\mathbf{r} + \mathbf{R}_i$, the phase of $\Psi_{\mathbf{k}}$ changes by $\Delta\theta = \mathbf{k}.\mathbf{R}_i$. Two points should be remarked:

1. The value of the phase at any one point is, as always, arbitrary, but $\Delta\theta$ for a translation $\mathbf{R}_i$ when the state is characterised by **k**, is strictly defined. The arbitrariness of the phase of course has no effect on any measurable quantity, but is of some consequence in the formalism of band theory (Section 3.4).
2. The change of phase is defined only apart from multiples of $2\pi$. Any vector $\mathbf{K}_n$ can be added to **k**, such that

$$\mathbf{K}_n.\mathbf{R}_i = 2\pi n \qquad (n = \pm 1, \pm 2, \ldots) \quad (2.10)$$

and the Bloch condition is still satisfied. This is the generalisation of what we found in the one-dimensional case; it is a point of the greatest importance (Section 2.9).

It is instructive to derive (2.9) again from a different point of view. Consider the translation operator $T_i$, being the operation of translating the lattice by $\mathbf{R}_i$:

$$T_i\Psi(\mathbf{r}) = \Psi(\mathbf{r} + \mathbf{R}_i) \tag{2.11}$$

The operators $T_i$ commute with each other and, since the potential is periodic, they commute with the Hamiltonian, so that their eigenvalues are constants of the motion. The eigenfunction of an electron in a crystal must then be a simultaneous eigenfunction of all $T_i$s. This means that, for every $T_i$,

$$T_i\Psi(\mathbf{r}) = C_i\Psi(\mathbf{r})$$

where $C_i$ is a number. Consider two translation operators $T_i$, $T_j$, then

$$T_jT_i\Psi(\mathbf{r}) = C_jC_i\Psi(\mathbf{r})$$

But the product $T_jT_i$ is itself a translation operator $T_m$, where

$$\mathbf{R}_m = \mathbf{R}_i + \mathbf{R}_j$$

and

$$T_m\Psi(\mathbf{r}) = C_m\Psi(\mathbf{r})$$

hence

$$C_m = C_iC_j$$

for any two lattice vectors $\mathbf{R}_i$, $\mathbf{R}_j$ such that

$$\mathbf{R}_i + \mathbf{R}_j = \mathbf{R}_m$$

Therefore the numbers $C_i$ must depend exponentially on $\mathbf{R}_i$

$$C(\mathbf{R}) = e^{\mathbf{b}.\mathbf{R}}$$

where the vector $\mathbf{b}$ characterises the particular $\Psi$. However, if $\mathbf{b}$ had a real part, $\Psi$ would tend exponentially to infinity as $\mathbf{R}$ increases; this is inadmissible, thus $\mathbf{b}$ must be of the form $i\mathbf{k}$, with $\mathbf{k}$ real and

$$T_i\Psi_{\mathbf{k}}(\mathbf{r}) = \Psi_{\mathbf{k}}(\mathbf{r} + \mathbf{R}_i) = e^{i\mathbf{k}.\mathbf{R}_i}\,\Psi_{\mathbf{k}}(\mathbf{r})$$

the Bloch condition, or Bloch theorem.

## 2.2 Bloch functions

Writing

$$u_{\mathbf{k}}(\mathbf{r}) = e^{-i\mathbf{k}.\mathbf{r}}\,\Psi_{\mathbf{k}}(\mathbf{r})$$

we have

$$u_{\mathbf{k}}(\mathbf{r} + \mathbf{R}_i) = \exp[-i\mathbf{k}.(\mathbf{r} + \mathbf{R}_i)]\Psi_{\mathbf{k}}(\mathbf{r} + \mathbf{R}_i) = e^{-i\mathbf{k}.\mathbf{r}}\,\Psi_{\mathbf{k}}(\mathbf{r}) = u_{\mathbf{k}}(\mathbf{r})$$

that is, $u_{\mathbf{k}}(\mathbf{r})$ has the periodicity of the lattice. We see then that

$$\Psi_{\mathbf{k}}(\mathbf{r}) = e^{i\mathbf{k}.\mathbf{r}}\,u_{\mathbf{k}}(\mathbf{r}) \tag{2.12}$$

The one-electron wavefunctions in perfect crystals are the product of a plane wave by a function which has the periodicity of the lattice. Such a wavefunction is called a Bloch function. We can guess that $u$ will behave near each nucleus much as an atomic wavefunction. Bloch functions are generally normalised in a unit cell of the crystal, or they may be normalised in unit volume, or in the parallelepiped used for the periodic boundary conditions. In this parallelepiped the Bloch functions form an orthogonal system.

It can be proved that the Coulomb and exchange integrals computed with Bloch functions have the correct periodicity. Hence the use of the Bloch theorem as a boundary condition is self-consistent. Note however that the Hartree-Fock equations may have other solutions, without the proper periodicity.

The vector **k** plays the role of a set of three quantum numbers. As for its physical meaning, we observe that if the potential is constant, and then it may be taken as zero, we have the free space solution of (2.1)

$$\Psi_{\mathbf{k}}(\mathbf{r}) = e^{i\mathbf{k}.\mathbf{r}}$$

with $E(\mathbf{k}) = \hbar^2k^2/2m$. $\hbar\mathbf{k}$ is in this case just the momentum of the electron. We shall see later that also in a crystal $\hbar\mathbf{k}$ has the properties of momentum as far as external forces are concerned. It is not however an eigenvalue of the operator $-i\hbar\nabla$ and is not equal to the average value of the momentum of the electron in state **k**. We have seen that **k** is a constant of the motion, that is the electron in state **k** is not scattered at all into other states by a perfect lattice. It is clear that $E$ is a function of **k** and may be considered as a continuous function, though for some values of **k** it may not be differentiable, and a particular dispersion relation $E(\mathbf{k})$ characterises each band. These relations are among the most important properties of a material. They enter in a fundamental way in any calculation of the electrical, optical, and thermal properties of the substance. Their determination is the fundamental problem of band theory, though of course for many problems a determination of the wavefunctions is also required.

Detailed knowledge of the band structure is becoming more and more essential in the design of semiconductor devices, such as tunnel diodes, Gunn oscillators, injection lasers, infrared detectors and other devices which depend on peculiar characteristics of the energy bands for their operation.

This is in contrast to the development of the transistor which required only the general concepts of the theory.

If now (2.12) is substituted into (2.1), the equation satisfied by $u_{\mathbf{k}}(\mathbf{r})$ is obtained

$$\nabla^2 u_{\mathbf{k}} + 2i\mathbf{k}.\nabla u_{\mathbf{k}} - k^2 u_k + \frac{2m_0}{\hbar^2}\Big[E(\mathbf{k}) - V(\mathbf{r})\Big]u_{\mathbf{k}} = 0 \qquad (2.13)$$

If we wish to write (2.13) as an eigenvalue equation, we introduce a **k**-dependent energy operator $H(\mathbf{k})$ operating on $u_{\mathbf{k}}(\mathbf{r})$

$$H(\mathbf{k})u_{\mathbf{k}}(\mathbf{r}) = \left[-\frac{\hbar^2}{2m_0}(\nabla + i\mathbf{k})^2 + V(\mathbf{r})\right]u_{\mathbf{k}}(\mathbf{r}) = E(\mathbf{k})u_{\mathbf{k}}(\mathbf{r}) \qquad (2.14)$$

and we note for future reference that

$$\text{grad}_{\mathbf{k}}\, H(\mathbf{k}) = -\frac{i\hbar^2}{m_0}(\nabla + i\mathbf{k}) \qquad (2.15)$$

Since $-i\hbar\nabla$ is the operator $\mathbf{p}$, we can write (2.13) as

$$\nabla^2 u_{\mathbf{k}} - \frac{2}{\hbar}\mathbf{k}.\mathbf{p}u_{\mathbf{k}} + \frac{2m_0}{\hbar^2}\left[E(\mathbf{k}) - \frac{\hbar^2 k^2}{2m_0} - V(\mathbf{r})\right]u_{\mathbf{k}} = 0 \qquad (2.16)$$

This equation is used to calculate $u$ corresponding to a certain **k** when $E$ and $u$ are known as a nearby $\mathbf{k}_0$; the term containing $\mathbf{k}.\mathbf{p}$ is treated as a perturbation ($\mathbf{k}.\mathbf{p}$ approximation).

If in (2.13) we change **k** in $-\mathbf{k}$, we have

$$\nabla^2 u_{-\mathbf{k}} - 2i\mathbf{k}.\nabla u_{-\mathbf{k}} - k^2 u_{-\mathbf{k}} + \frac{2m_0}{\hbar^2}\Big[E(-\mathbf{k}) - V(\mathbf{r})\Big]u_{-\mathbf{k}} = 0 \qquad (2.17)$$

By comparing (2.17) with the complex conjugate of (2.13) we see that $E$ must be an even function of **k** and $u_{-\mathbf{k}}$ must be a multiple of $u_{\mathbf{k}}^*$

$$E(-\mathbf{k}) = E(\mathbf{k})$$

$$u_{-\mathbf{k}}(\mathbf{r}) = Cu_{\mathbf{k}}^*(\mathbf{r}) \qquad (2.18)$$

$C$ being a complex constant of modulus one. This constant can be adjusted to be one, or any chosen phase factor, by adjusting the arbitrary phase of $u_{\mathbf{k}}(\mathbf{r})$. In a similar way it can be proved that, provided the crystal has a centre of inversion at the origin,

$$u_{\mathbf{k}}^*(-\mathbf{r}) = C' u_{\mathbf{k}}(\mathbf{r}) \qquad (2.19)$$

$C'$ having the same properties as $C$. The symmetry expressed by these relations does not follow from crystal symmetry. It is known as time reversal symmetry. It may be considered as the analogue, for electron waves, of the optical principle of reversibility of the light path.

## 2.3 A one-dimensional example

In order to see how energy bands and forbidden energy gaps arise as a consequence of the periodicity conditions, consider the one-dimensional problem of the determination of the electronic energy levels in the potential

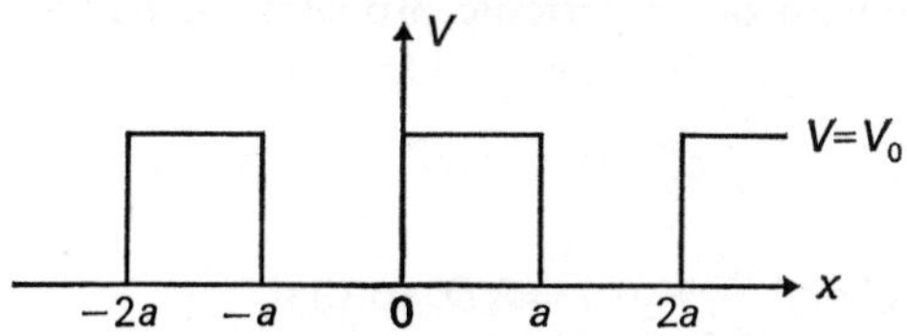


Fig. 2.1. Potential in one-dimensional example.

of Fig. 2.1. The period of this structure is $2a$, and thus the Bloch conditions are

$$\Psi_k(x + 2a) = e^{2iak}\,\Psi_k(x)$$

$$\frac{d}{dx}\Psi_k(x + 2a) = e^{2iak}\,\frac{d}{dx}\Psi_k(x) \tag{2.20}$$

The 'wells' of the potential may be thought to represent the atoms of the crystal, and the 'hills' the interatomic potential barriers. According to classical mechanics, if the energy $E$ of the electron is less than $V_0$, the electron remains confined in the well. If $E > V_0$, the velocity is $\left(\dfrac{2E}{m_0}\right)^{\frac{1}{2}}$ in a well and $\left(2\dfrac{E - V_0}{m_0}\right)^{\frac{1}{2}}$ on a hill. Classically the system is a high-pass filter. However, according to quantum mechanics it transmits a series of bands. $E$ can be less than $V_0$ because of the tunnelling effect through potential barriers.

The solution of Schrödinger's equation for $-a \leqq x \leqq 0$ is

$$\Psi_1 = A\cos q_0 x + B\sin q_0 x$$

with

$$q_0^2 = \frac{2m_0E}{\hbar^2} \tag{2.21}$$

$A$, $B$ being two constants to be determined by the continuity and periodicity conditions. Similarly, for $0 \leqq x \leqq a$,

$$\Psi_2 = C \cos qx + D \sin qx \tag{2.22}$$

with

$$q^2 = \frac{2m(E - V_0)}{\hbar^2}$$

$q$ is imaginary if $E < V_0$. As boundary conditions we take

1. $\Psi$ and $\frac{d}{dx}\Psi$ continuous at $x = 0$;
2. $\Psi(+a) = e^{2iak}\,\Psi(-a)$; $\quad \frac{d}{dx}\Psi(+a) = e^{2iak}\frac{d}{dx}\Psi(-a)$.

Note the analogy with other periodic problems of physics, for instance, in electromagnetic theory, with the problem of propagation on a periodic transmission line: instead of current and potential, we have $\Psi$ and $\frac{d}{dx}\Psi$.

Now $\Psi_2(0) = C$ and $\frac{d}{dx}\Psi_2(0) = qD$, so that

$$\Psi_2(+a) = \cos qa\Psi_2(0) + \frac{1}{q}\sin qa\frac{d}{dx}\Psi_2(0)$$

$$\frac{d}{dx}\Psi_2(+a) = -q\sin qa\Psi_2(0) + \cos qa\frac{d}{dx}\Psi_2(0)$$

or, in matrix form,

$$\begin{vmatrix} \Psi_2(a) \\ \frac{d}{dx}\Psi_2(a) \end{vmatrix} = \begin{vmatrix} \cos qa & \frac{1}{q}\sin qa \\ -q\sin qa & \cos qa \end{vmatrix} \begin{vmatrix} \Psi_2(0) \\ \frac{d}{dx}\Psi_2(0) \end{vmatrix}$$

The determinant of the linear transformation (scattering matrix) has the value unity. Similarly

$$\begin{vmatrix} \Psi_1(0) \\ \frac{d}{dx}\Psi_1(0) \end{vmatrix} = \begin{vmatrix} \cos q_0a & \frac{1}{q_0}\sin q_0a \\ -q_0\sin q_0a & \cos q_0a \end{vmatrix} \begin{vmatrix} \Psi_1(-a) \\ \frac{d}{dx}\Psi_1(-a) \end{vmatrix}$$

Now $\Psi_1(0) = \Psi_2(0)$, and $\frac{d}{dx}\Psi_1(0) = \frac{d}{dx}\Psi_2(0)$, so that

$$\begin{vmatrix} \Psi_2(a) \\ \frac{d}{dx}\Psi_2(a) \end{vmatrix} = \begin{vmatrix} b_{11} & b_{12} \\ b_{21} & b_{22} \end{vmatrix} \begin{vmatrix} \Psi_1(-a) \\ \frac{d}{dx}\Psi_1(-a) \end{vmatrix}$$

with $b_{11} = \cos qa \cos q_0a - \dfrac{q_0}{q} \sin qa \sin q_0a$ and

$$b_{22} = -\frac{q}{q_0} \sin qa \sin q_0a + \cos qa \cos q_0a, \text{ and Det } b_{ij} = 1 .$$

However we must have

$$\begin{vmatrix} \Psi_2(a) \\ \dfrac{d}{dx}\Psi_2(a) \end{vmatrix} = e^{2ika} \begin{vmatrix} \Psi_1(-a) \\ \dfrac{d}{dx}\Psi_1(-a) \end{vmatrix}$$

so that

$$\begin{vmatrix} b_{11} - e^{2iak} & b_{12} \\ b_{21} & b_{22} - e^{2iak} \end{vmatrix} = 0$$

with the solution (since Det $b_{ij} = 1$)

$$e^{2iak} = \tfrac{1}{2}(b_{11} + b_{22}) \pm i[1 - \tfrac{1}{4}(b_{11} + b_{22})^2]^{\frac{1}{2}}$$

or

$$\cos 2ka = \tfrac{1}{2}(b_{11} + b_{22})$$
$$= \cos qa \cos q_0a - \frac{q^2 + q_0^2}{2qq_0} \sin qa \sin q_0a \tag{2.23}$$

Since $q$ and $q_0$ contain the energy $E$, the last equation is the dispersion relation $E = E(k)$ in an implicit form. It is clear that, since $k$ must be real, the problem has a physically acceptable solution only if the absolute value of the right hand side does not exceed unity. Allowed and forbidden bands of energy then arise.

If $E < V_0$, $q$ is imaginary $q = ip$; then

$$\cos 2ka = \cosh pa \cos q_0a - \frac{q_0^2 - p^2}{2pq_0} \sinh pa \sin q_0a \tag{2.24}$$

If $V_0$ is very large and $E \ll V_0$ (deep levels), the hyperbolic functions are very large, and practically equal. Equation (2.24) becomes approximately, since $p \gg q_0$,

$$\cos 2ka = \cosh pa \left[ \cos q_0a + \frac{p}{2q_0} \sin q_0a \right] \tag{2.25}$$

This can be satisfied only in a narrow interval of energies for which $\sin q_0a$ is negative and very small (if $\cos q_0a$ is positive). The condition $\sin q_0a = 0$ gives

$$E_n = n^2 \frac{\pi^2\hbar^2}{2m_0a^2} \tag{2.26}$$

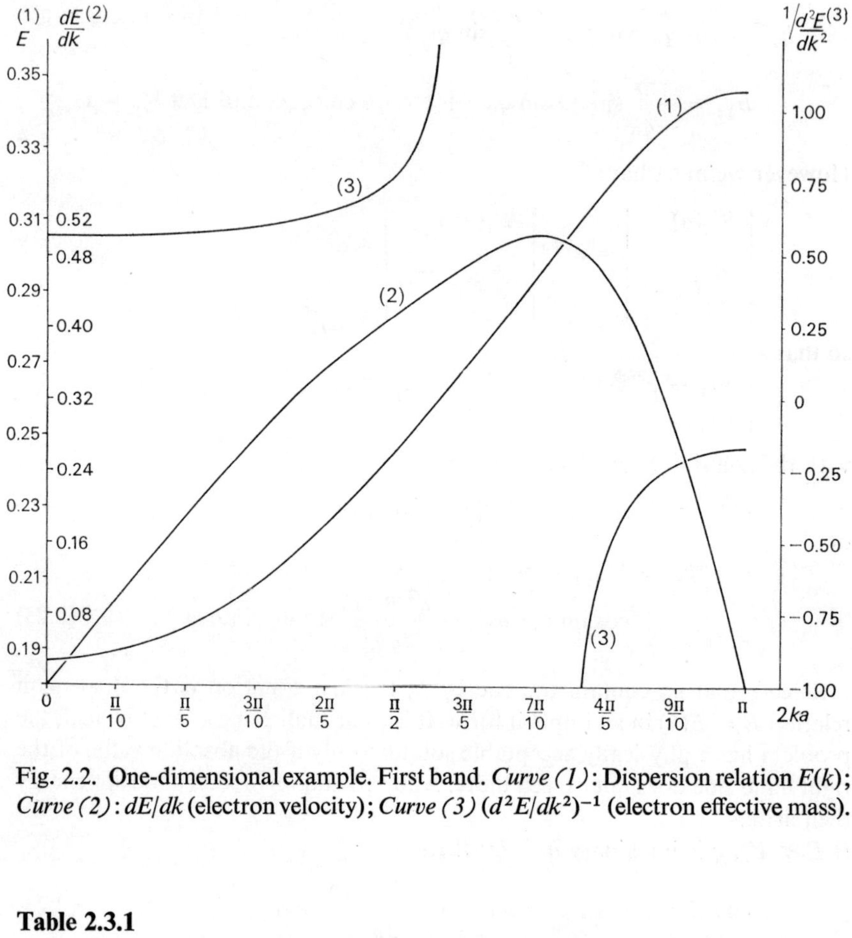


Fig. 2.2. One-dimensional example. First band. *Curve (1)*: Dispersion relation $E(k)$; *Curve (2)*: $dE/dk$ (electron velocity); *Curve (3)* $(d^2E/dk^2)^{-1}$ (electron effective mass).

**Table 2.3.1**

| $E$ | cos $2ka$ | $k$ | | $E$ | cos $2ka$ | $k$ | |
|---|---|---|---|---|---|---|---|
| 0 | 7·39 | imaginary | | 1·31–1·35 | >1 | imaginary | |
| 0·16 | 1·58 | imaginary | | 1·351 | 1 | 0 | third band |
| 0·1863 | 1 | 0 | first band | 2·65 | −1 | $\pm\pi/6$ | |
| 0·2444 | 0 | $\pm\pi/12$ | | 2·65–2·74 | <−1 | imaginary | |
| 0·3461 | −1 | $\pm\pi/6$ | | 2·74 | −1 | $\pm\pi/6$ | fourth band |
| 0·4444 | −1·325 | imaginary | | 4·61 | 1 | 0 | |
| 0·6271 | −1 | $\pm\pi/6$ | second band | . . . | . . . | . . . | |
| 0·8644 | 0 | $\pm\pi/12$ | | | | | |
| 1·3077 | 1 | 0 | | | | | |

These are the well known eigenvalues for a potential well of infinite depth, corresponding to deep atomic levels. Such levels are thus only slightly broadened and displaced when the atoms form a crystal.

A numerical example is given in Table 2.3.1 which shows the various bands for $2a = 6$ Bohr radii (interatomic spacing of about 3 Å) and $V_0 = 4/9$ Rydberg (about 6 eV). The energy gaps become smaller and smaller for higher energies. The smallest value of $k$ is given throughout, but $E$ is periodic in $k$ (with period $\pi/3$).

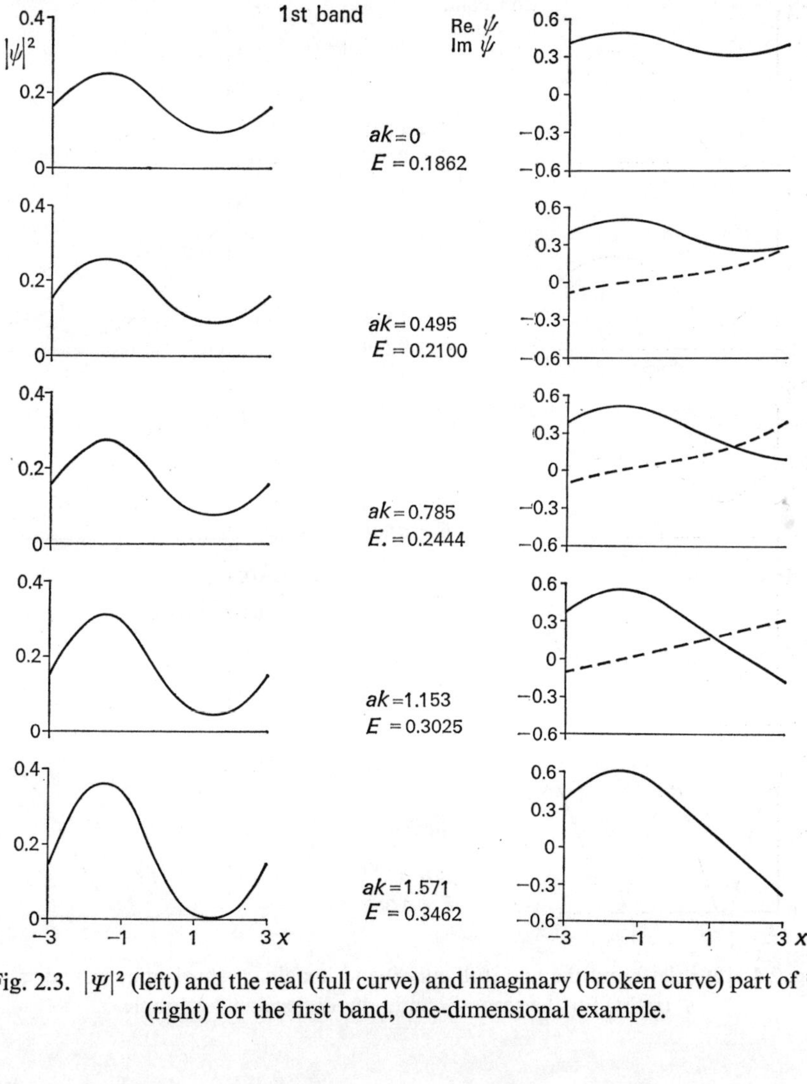


Fig. 2.3. $|\Psi|^2$ (left) and the real (full curve) and imaginary (broken curve) part of $\Psi$ (right) for the first band, one-dimensional example.

Fig. 2.2. gives the $E$–$k$ curve for the first band. It gives also $dE/dk$, which, as we shall see, is proportional to the velocity of the electron, and $(d^2E/dk^2)^{-1}$, which, as we shall see, is proportional to its effective mass.

Fig. 2.3 and Fig. 2.4 give $|\Psi|^2$, normalised in the unit cell, and the real and imaginary parts of $\Psi$ for five states respectively in the first and in the second band. The phase has been chosen so that $\Psi$ is either real or purely imaginary at the centre $x_0$ of the potential well. Other choices can be made; for instance,

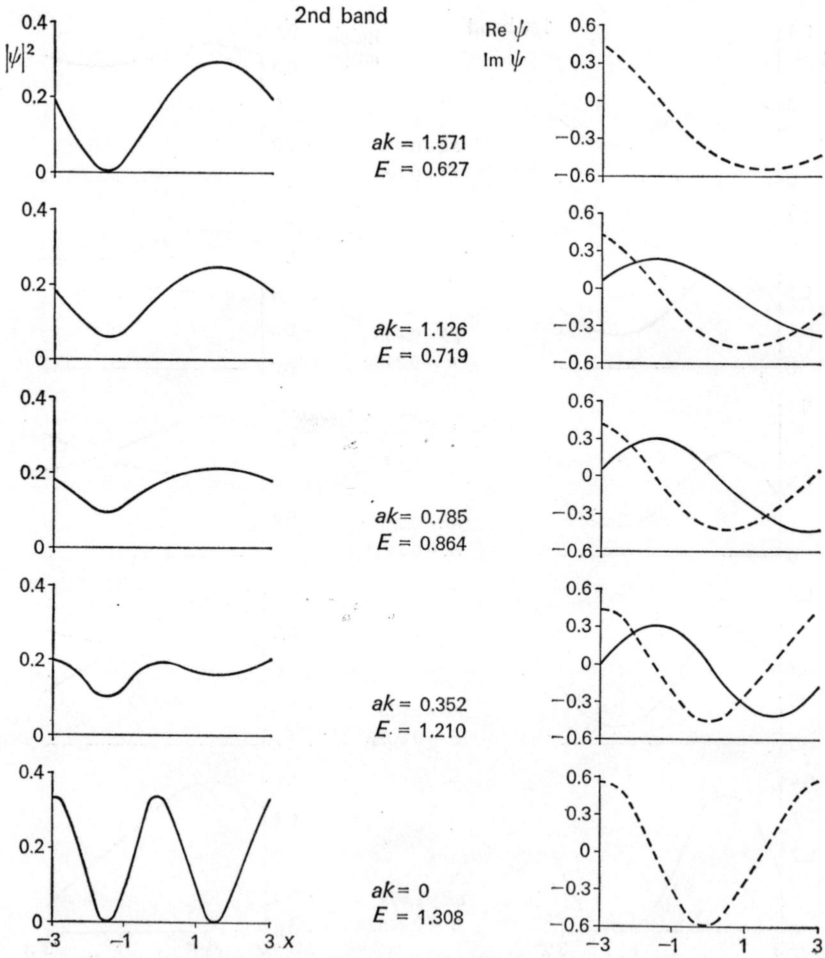


Fig. 2.4. $|\Psi|^2$ (left) and the real (full curve) and imaginary (broken curve) parts of $\Psi$ (right) for the second band, one-dimensional example.

$\Psi$ can be taken as real at $x = 0$. The choice made is convenient because equation (2.19) shows that then the real part of $\Psi$ is an even function of $x - x_0$ and the imaginary part an odd function or vice versa. Of course, $|\Psi|^2$ is independent of the choice of the phase. At the band edges $\Psi$ can be taken either as real or as purely imaginary: these standing waves are due to the Bragg reflections.

## 2.4 Variational method

In the example of Section 2.3 it was possible to find the dispersion relation in a closed form. However this is usually impossible, and the almost universal procedure is to derive the energy values for any chosen **k** by means of the variational method; that is, one calculates the expectation values $W$ of the energy with a normalised trial function $\Psi_t$ (depending on **k**, though the suffix **k** is omitted in the following for simplicity). Then

$$W = \int \Psi_t^* H \Psi_t \, d\mathbf{r} \tag{2.27}$$

where $H$ is the Hamiltonian operator, provides an upper limit to the lowest eigenvalue $E_0$. As a trial function, a linear combination is taken of $n$ functions $\chi_i$ (basis functions) chosen from a complete orthonormal set

$$\Psi_t = \sum_{i=1}^{n} a_i \chi_i \tag{2.28}$$

where $n$ depends on the accuracy required and on the computational facilities. Substituting (2.28) in (2.27), and using the notation

$$H_{ij} = \int \chi_i^* H \chi_j \, d\mathbf{r}$$

(matrix element of the Hamiltonian between the $i$ and $j$ basis functions), we have

$$W = \sum_{i=1}^{n} \sum_{j=1}^{n} a_i^* a_j H_{ij} \tag{2.29}$$

an expression which is to be minimised with respect to the coefficients $a_i$, subject however to the normalisation condition

$$\sum_{i=1}^{n} a_i^* a_i = 1$$

One minimises then, without conditions, the expression

$$\Sigma_i \Sigma_j \, a_i^* a_j H_{ij} - \lambda[\Sigma_i \, a_i^* a_i - 1] \tag{2.30}$$

where $\lambda$ is a Lagrangian multiplier. Differentiation with respect to $a_i^*$ gives the system of equations

$$\sum_{j=1}^{n} a_j H_{ij} = \lambda a_i \qquad i = 1, 2, \ldots n \tag{2.31}$$

which have non-zero solutions only if

$$\begin{vmatrix} H_{11} - \lambda & H_{12} & \ldots & H_{1n} \\ H_{21} & H_{22} - \lambda & \ldots & H_{2n} \\ \cdot & \cdot & \ldots & \cdot \\ H_{n1} & H_{n2} & \ldots & H_{nn} - \lambda \end{vmatrix} = 0 \tag{2.32}$$

This determinantal equation gives $n$ roots for $\lambda$, and for each of them the coefficients $a_i$ can be determined. Inserting (2.31) in (2.29) we obtain for any root

$$W = \Sigma_i \, a_i^* \lambda a_i = \lambda \tag{2.33}$$

Thus the lowest root of (2.32) provides the minimum value of $W$, which is an upper limit to $E_0$. The other roots approximate energies of higher bands.

It should be kept in mind that a first order error in the trial function produces a second order error in the energy. Fairly satisfactory energy values can be obtained with somewhat imperfect wavefunctions, but there is no guarantee that, even if the energy levels are correct, other quantities calculated with the imperfect wavefunctions will be in agreement with experiment.

The different methods of band structure calculation differ essentially in the choice of the orthonormal system of basis functions (Section 5.7).

## 2.5 Conductors and insulators

The fact that an energy band whose every state is occupied (by two electrons with opposite spins) cannot carry an electric current is the basis of the well known quantum-mechanical explanation of the difference between insulators and conductors: insulators have only completely full bands. However, if the energy difference between the uppermost completely filled band and the next empty band is not much larger than $kT$, where $k$ is Boltzmann's constant and $T$ is the absolute temperature, thermal agitation can raise electrons to the empty (conduction) band, leaving empty places in the full (valence) band, and we have then an intrinsic semiconductor, with mobile electrons and holes, whose conductivity increases as $T$ increases. If there are incompletely filled bands, or overlapping bands, higher energy levels with infinitesimally higher energy are available to the uppermost electrons, and the substance is a conductor.

It is interesting to observe that, according to this picture, an array of hydrogen-like atoms, with a half-filled band, ought to be a metal even at large lattice spacing (though the effective mass would become very large), which is against common experience and common sense. Mott (1961) has argued that in such cases there must be a sharp transition between the conducting and non-conducting states. In fact, if there are few carriers, a hole and an electron, even if formed, will remain bound to each other in a sort of positronium atom, which is called an exciton, and all these states are localised. But, if there is a considerable concentration of free carriers, the Coulomb field is replaced by a screened one, and the bound states disappear. This occurs when the screening distance becomes of the same order as the radius of the orbit. Now the screening radius, if the density of the free electrons is $n$, is

$$r_s \simeq \frac{\hbar}{2em_0^{\frac{1}{2}}n^{\frac{1}{6}}} \tag{2.34}$$

(Section 4.2), while the radius of the orbit is of the order of the radius of the first orbit of hydrogen

$$r_H \simeq \frac{\hbar^2}{m_0 e^2} \tag{2.35}$$

thus the transition can be expected when $n^{\frac{1}{3}} \simeq \dfrac{1}{4r_H}$, $n \simeq 10^{19}\ \text{cm}^{-3}$.

Of course this simple argument cannot yield quantitatively correct results, but it shows that the number of free carriers should vary with lattice distance $a$ as indicated in Fig. 2.5. Mathematically, at the transition, one should switch

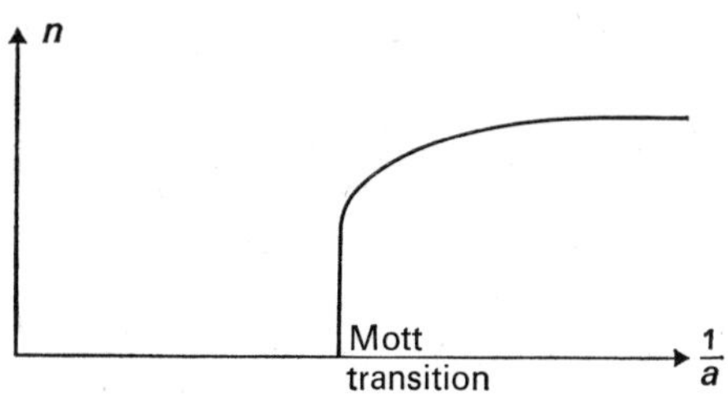


Fig. 2.5. Metal-insulator transition according to Mott.

from a description in terms of Bloch functions to a description in terms of localised orbitals.

The Mott transition has been observed in certain oxides which become conducting suddenly above a certain temperature. It may possibly explain the property of impurity levels in semiconductors to merge in an impurity band*.

* It has been argued that the sudden large increase in conductivity of certain glasses at a critical electric field is an allied phenomenon.

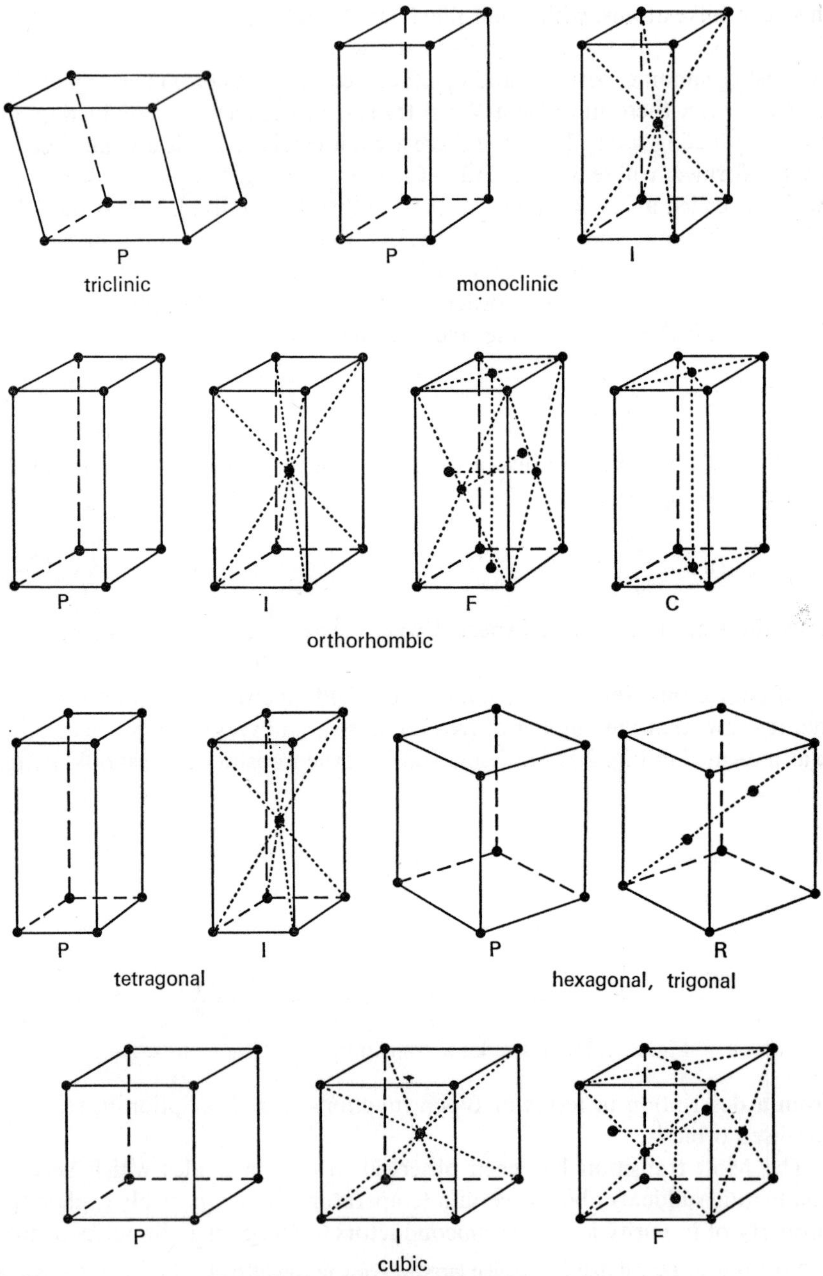


Fig. 2.6. The 14 Bravais lattices.

## 2.6 Bravais lattices

A crystal lattice has infinite periods, that is, for a lattice of infinite extension there are infinite displacements that bring the lattice to a completely equivalent position. Choose periods $\mathbf{a}_1$, $\mathbf{a}_2$, $\mathbf{a}_3$ such that $\mathbf{a}_1$ is the shortest period in the lattice (or one of several equal shortest ones), $\mathbf{a}_2$ is the shortest not parallel to $\mathbf{a}_1$, and $\mathbf{a}_3$ is the shortest not co-planar to $\mathbf{a}_1$ and $\mathbf{a}_2$. Then any period $\mathbf{R}$ of the lattice (primitive translation) is given by $\mathbf{R} = l\mathbf{a}_1 + m\mathbf{a}_2 + n\mathbf{a}_3$, with $l, m, n$ integers.

The lattice so defined is called the Bravais lattice of the crystal to which it belongs. The crystal is completely described by the contents of one parallelepiped of sides $\mathbf{a}_1$, $\mathbf{a}_2$, $\mathbf{a}_3$. This unit is called a primitive cell of the crystal and it may contain any number of atoms. The volume $\Omega$ of the primitive cell is $\mathbf{a}_1 \cdot \mathbf{a}_2 \times \mathbf{a}_3$: this cell is the smallest unit whose periodic repetition builds up the lattice. Any other cell whose translation generates the lattice is called a unit cell.

There are 14 different Bravais lattices (Fig. 2.6) which are classified into seven systems:

*triclinic*, in which $\mathbf{a}_1$, $\mathbf{a}_2$, $\mathbf{a}_3$ have different lengths and no two of them are perpendicular;

*monoclinic*, in which the three vectors still have different lengths, but $\mathbf{a}_3$ is perpendicular to the plane of $\mathbf{a}_1$ and $\mathbf{a}_2$;

*orthorhombic*, in which the three vectors are mutually perpendicular, but still have different lengths;

*tetragonal*, in which the vectors are mutually perpendicular and $\mathbf{a}_1 = \mathbf{a}_2$;

*trigonal*, in which the three vectors have equal lengths and the angles between $\mathbf{a}_1$, $\mathbf{a}_2$, between $\mathbf{a}_2$, $\mathbf{a}_3$, and between $\mathbf{a}_3$, $\mathbf{a}_1$ are equal but not 90°;

*hexagonal*, in which $\mathbf{a}_1$ and $\mathbf{a}_2$ are equal and form an angle of 120°, while $\mathbf{a}_3$ is perpendicular to the plane of $\mathbf{a}_1$ and $\mathbf{a}_2$;

*cubic*, in which all vectors are equal and perpendicular.

However, in some cases the vectors must be chosen differently from the way just described to get primitive cells which are the smallest periodic units; in any case these cells must contain one lattice point each, must be identical and must fill all space.

There are three Bravais lattices in the cubic system. The simple cubic lattice (s.c.l.) ($P$) corresponds to the above definition. The body-centred cubic lattice (b.c.c.l.) ($I$) has lattice points at

$$x = n_1 a, \qquad y = n_2 a, \qquad z = n_3 a$$

and also at

$$x = (n_1 + \tfrac{1}{2})a, \qquad y = (n_2 + \tfrac{1}{2})a, \qquad z = (n_3 + \tfrac{1}{2})a$$

The cubic cells are not primitive, they contain two equivalent lattice points. It is then possible to choose a primitive unit cell of volume $\frac{1}{2}a^3$ and to find

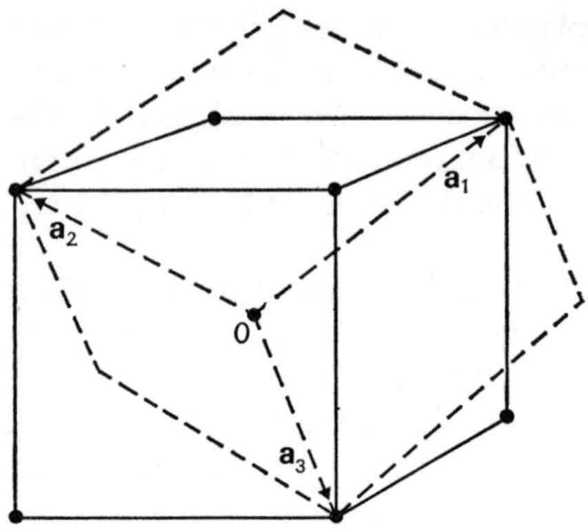


Fig. 2.7. Primitive cell of body-centred cubic lattice.

vectors $\mathbf{a}_1$, $\mathbf{a}_2$, $\mathbf{a}_3$ appropriate to it. Fig. 2.7 shows the simplest way of choosing these vectors:

$$\begin{aligned}\mathbf{a}_1 &= \tfrac{1}{2}a(-\mathbf{i}+\mathbf{j}+\mathbf{k})\\ \mathbf{a}_2 &= \tfrac{1}{2}a(\mathbf{i}-\mathbf{j}+\mathbf{k})\\ \mathbf{a}_3 &= \tfrac{1}{2}a(\mathbf{i}+\mathbf{j}-\mathbf{k})\end{aligned}$$

The primitive unit cell may be taken as the parallelepiped whose sides are parallel to $\mathbf{a}_1$, $\mathbf{a}_2$, $\mathbf{a}_3$, but this cell has not the complete cubic symmetry. The cell which has this summetry is the Wigner-Seitz cell.

For any crystal, the Wigner-Seitz cell associated with a lattice point $\mathbf{R}$ is defined as the locus of all points that are closer to $\mathbf{R}$ than to any other lattice point. For the b.c.c.l. the Wigner-Seitz cell is a truncated octahedron (Fig. 2.8).

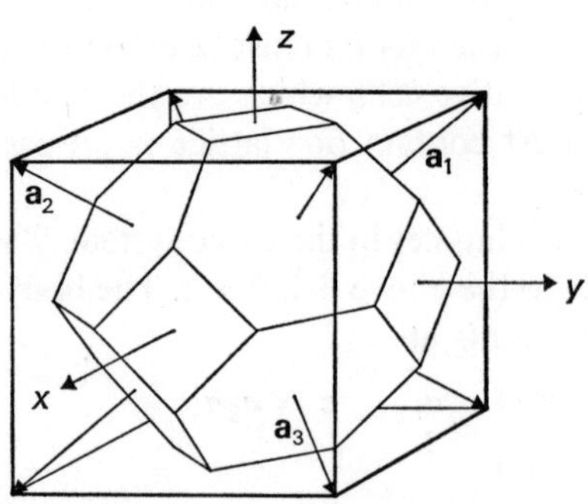


Fig. 2.8. Wigner-Seitz cell of the body-centred cubic lattice.

The f.c.c.l. (F) has lattice points at

$$\begin{array}{lll} x = n_1 a, & y = n_2 a, & z = n_3 a \\ x = (n_1 + \tfrac{1}{2})a, & y = (n_2 + \tfrac{1}{2})a, & z = n_3 a \\ x = (n_1 + \tfrac{1}{2})a, & y = n_2 a, & z = (n_3 + \tfrac{1}{2})a \\ x = n_1 a, & y = (n_2 + \tfrac{1}{2})a, & z = (n_3 + \tfrac{1}{2})a \end{array}$$

The cubic cell contains 4 lattice points. One can choose

$$\mathbf{a}_1 = \tfrac{1}{2}a(\mathbf{j} + \mathbf{k})$$
$$\mathbf{a}_2 = \tfrac{1}{2}a(\mathbf{k} + \mathbf{i})$$
$$\mathbf{a}_3 = \tfrac{1}{2}a(\mathbf{i} + \mathbf{j})$$

The primitive cell then has volume $a^3/4$. It is illustrated in Fig. 2.9. The Wigner-Seitz cell for this structure is a rhombododecahedron (Fig. 2.10).

The hexagonal system has only the simple hexagonal Bravais lattice ($P$).

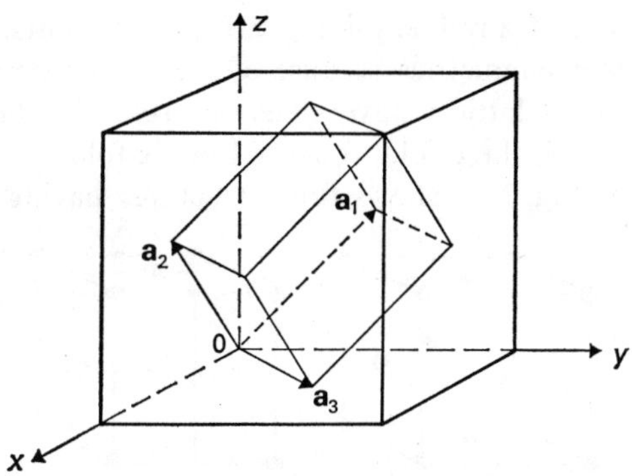


Fig. 2.9. Primitive cell of the face-centred cubic lattice.

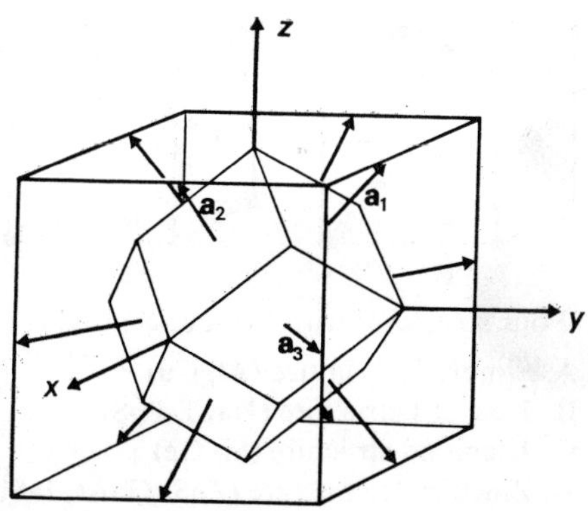


Fig. 2.10. Wigner-Seitz cell of the face-centred cubic lattice.

Three of the cells illustrated in Fig. 2.6 must be fitted together to obtain the hexagonal symmetry. The Wigner-Seitz cell is a hexagonal prism. In the trigonal system there is again the hexagonal lattice *P* and the rhombohedral lattice *R*. The tetragonal system has the simple tetragonal (*P*) and the body-centred tetragonal (I) Bravais lattices. The orthorhombic system includes four Bravais lattices: the simple (*P*), the face-centred (*F*) and the body-centred (I) orthorhombic lattices and the one-face-centred orthorhombic lattice (*C*), in which the centres of one set of faces are equivalent to the corners of the cell, but not the centres of the other two sets of faces. The monoclinic system has the simple (*P*) and the one-face-centred (*I*) monoclinic lattices. The triclinic system has only a simple lattice.

In all cases the primitive cell may be filled in a number of different ways, so that the 14 Bravais lattices become diversified into 230 space groups. Each of these groups can be the symmetry type of several different crystal structures.

A Bravais lattice has inversion symmetry about every lattice point, and also about the mid-point of any line joining two lattice points. However, many structures lack these symmetries because of the distribution of the atoms in the unit cells. Bravais lattices have generally axes of rotational symmetry which can be two-fold, three-fold, four-fold or six-fold.

As an example, Fig. 2.11 shows four structures having the face-centred

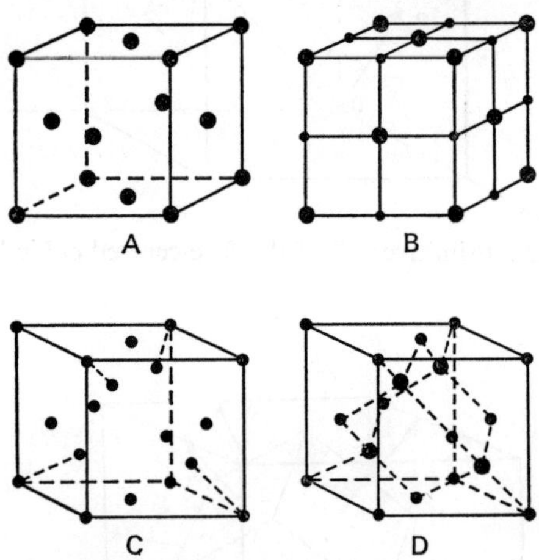


Fig. 2.11. Four structures with the face-centred cubic lattice.

A. Simple f.c.c. lattice (Al, Cu)
B. Rocksalt structure (NaCl, PbS)
C. Diamond structure (Si, Ge)
D. Zincblende structure (ZnS, GaAs, InSb)

Note tetrahedral bonds in C, D.

cubic Bravais lattice but four different space groups: the monatomic face-centred cubic structure (aluminium), the sodium chloride structure, the diamond structure (silicon, germanium) and the zincblende structure (GaAs, InSb). In the last three structures there are two atoms per unit cell. When there is more than one atom per unit cell, we speak of a lattice with a basis. We must assign the vectors, within each cell, that produce the lattice. Thus for the diamond structure there is the single basis vector $a(\frac{1}{4},\frac{1}{4},\frac{1}{4})$. The crystals are made up of two interpenetrating f.c.c. lattices. The diamond structure has inversion symmetry about the mid-point of any line joining two nearest atoms, but this symmetry is lost in the zincblende structure because the two atoms are different. This fact leads to certain small fundamental differences between the band structure of elements with the diamond structure, such as Si, Ge, and compounds with the zincblende structure, such as GaAs, InSb (Section 6.5).

Similarly, some of the rotational symmetries of the Bravais lattice may not

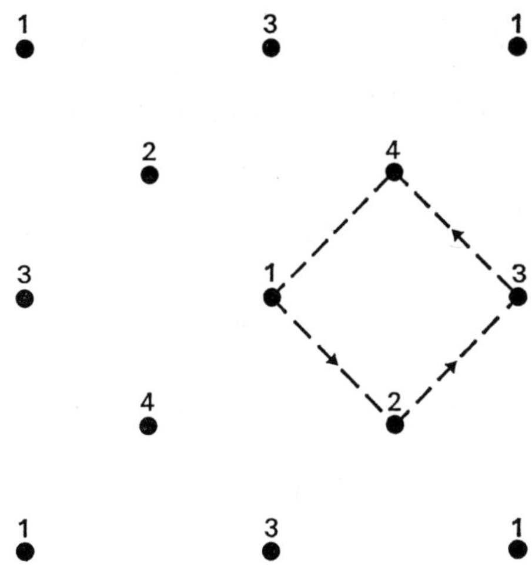


Fig. 2.12. Projection of the contents of the unit cell of the diamond structure (Fig. 2.11 C) on a face of the elementary cube. The numbers indicate the four layers; the fifth layer is the same as the first. The dashed square shows how the cubic fourfold symmetry has become a screw symmetry.

be present in the crystal, or retained in a modified form, for example a rotation about an axis of symmetry of the Bravais lattice must be combined with a translation in the crystal, by a fraction of a lattice displacement. Thus in the diamond structure the normal cubic fourfold symmetry becomes a screw symmetry (Fig. 2.12): the screw axes are perpendicular and pass through the centre of the faces of the cube. The presence of screw axes, or glide planes, has important effects on the band structure.

## 2.7 Reciprocal space

For the study of the $E(\mathbf{k})$ relations it is convenient to establish a space for the crystal momentum $\mathbf{k}$. Given a lattice with basis vectors $\mathbf{a}_1$, $\mathbf{a}_2$, $\mathbf{a}_3$, one defines as its reciprocal lattice that generated by three basis vectors, said to be in reciprocal space, $\mathbf{b}_1$, $\mathbf{b}_2$, $\mathbf{b}_3$, satisfying

$$\mathbf{a}_i.\mathbf{b}_j = 2\pi\delta_{ij} \tag{2.36}$$

If the right hand sides in (2.36) are taken to be 1 instead of $2\pi$, we have the so-called $\mathbf{k}$-space; if they are taken to be $2\pi\hbar$, we have momentum space, though the three terms are often used synonymously. It is clear that the basis vectors of reciprocal space are perpendicular to the basis planes of the direct lattice. Thus

$$\mathbf{b}_1 = C\mathbf{a}_2 \times \mathbf{a}_3$$

To find the constant $C$ we multiply scalarly by $\mathbf{a}_1$ and use (2.36) obtaining

$$C = \frac{2\pi}{\mathbf{a}_1.\mathbf{a}_2 \times \mathbf{a}_3}$$

But

$$\mathbf{a}_1.\mathbf{a}_2 \times \mathbf{a}_3 = \Omega$$

the volume of the primitive cell of the direct lattice, thus

$$\mathbf{b}_1 = \frac{2\pi}{\Omega}\mathbf{a}_2 \times \mathbf{a}_3$$

and similarly

$$\mathbf{b}_2 = \frac{2\pi}{\Omega}\mathbf{a}_3 \times \mathbf{a}_1 \qquad \mathbf{b}_3 = \frac{2\pi}{\Omega}\mathbf{a}_1 \times \mathbf{a}_2 \tag{2.37}$$

Any lattice vector, $\mathbf{K}_n$ (primitive translation), of the reciprocal lattice is given by

$$\mathbf{K}_n = p\mathbf{b}_1 + q\mathbf{b}_2 + r\mathbf{b}_3 \qquad (p, q, r \text{ integers}) \tag{2.38}$$

If $\mathbf{R}_i$ is any direct lattice vector,

$$\mathbf{R}_i = l\mathbf{a}_1 + m\mathbf{a}_2 + n\mathbf{a}_3$$

we have from (2.36)

$$\mathbf{K}_n.\mathbf{R}_i = 2\pi(pl + qm + rn) = 2\pi g \; (g \text{ an integer})$$

so that the $\mathbf{K}_n$ are the vectors considered in (2.10) such that

$$\exp(i\mathbf{K}_n.\mathbf{R}_i) = 1 \tag{2.39}$$

These vectors are introduced in the most natural way by expanding the potential $V(\mathbf{r})$, or any other function with the periodicity of the lattice, in a Fourier series

$$V(\mathbf{r}) = \Sigma_{\mathbf{K}_n} V(\mathbf{K}_n)\, e^{i\mathbf{K}_n.\mathbf{r}} \tag{2.40}$$

the $\mathbf{K}_n$ are the 'frequencies' for the basic periodic functions $e^{i\mathbf{K}_n.\mathbf{r}}$. It is immediately verified that, because of (2.39), these functions have the periodicity of the lattice.

The volume of the primitive cell of the reciprocal lattice, $\mathbf{b}_1.\mathbf{b}_2 \times \mathbf{b}_3$, is $8\pi^3$ divided by the volume $\Omega$ of the primitive cell of the direct lattice. In fact,

$$\begin{aligned}\mathbf{b}_2 \times \mathbf{b}_3 &= \frac{4\pi^2}{\Omega^2}(\mathbf{a}_3 \times \mathbf{a}_1) \times (\mathbf{a}_1 \times \mathbf{a}_2)\\ &= \frac{4\pi^2}{\Omega^2}\{\mathbf{a}_1[\mathbf{a}_2.(\mathbf{a}_3 \times \mathbf{a}_1)] - \mathbf{a}_2[\mathbf{a}_1.(\mathbf{a}_3 \times \mathbf{a}_1)]\}\\ &= \frac{4\pi^2}{\Omega}\mathbf{a}_1\end{aligned}$$

and thus

$$\begin{aligned}\mathbf{b}_1.\mathbf{b}_2 \times \mathbf{b}_3 &= \frac{4\pi^2}{\Omega}\mathbf{a}_1.\mathbf{b}_1\\ &= \frac{8\pi^3}{\Omega}\end{aligned} \tag{2.41}$$

From the definition of $\mathbf{b}_1$, $\mathbf{b}_2$, $\mathbf{b}_3$, it is easily seen that the reciprocal lattice of a simple cubic lattice is also simple cubic. As for the reciprocal lattice of a face-centred cubic lattice, consider the primitive vectors of Fig. 2.9

$$\mathbf{a}_1 = \tfrac{1}{2}a(\mathbf{j} + \mathbf{k}), \qquad \mathbf{a}_2 = \tfrac{1}{2}a(\mathbf{k} + \mathbf{i}), \qquad \mathbf{a}_3 = \tfrac{1}{2}a(\mathbf{i} + \mathbf{j})$$

Then

$$\mathbf{a}_2 \times \mathbf{a}_3 = \frac{a^2}{4}(-\mathbf{i} + \mathbf{j} + \mathbf{k})$$

and since

$$\Omega = \frac{a^3}{4}$$

$$\mathbf{b}_1 = \frac{2\pi}{\Omega}\mathbf{a}_2 \times \mathbf{a}_3 = \frac{2\pi}{a}(-\mathbf{i} + \mathbf{j} + \mathbf{k})$$

Similarly

$$\mathbf{b}_2 = \frac{2\pi}{a}(\mathbf{i} - \mathbf{j} + \mathbf{k}), \qquad \mathbf{b}_3 = \frac{2\pi}{a}(\mathbf{i} + \mathbf{j} - \mathbf{k})$$

thus the reciprocal lattice is body-centred cubic (Fig. 2.7). The side of the elementary cube is $4\pi/a$. Conversely, the reciprocal lattice of a b.c.c. lattice is f.c.c. Ten of the 14 Bravais lattices are their own reciprocals.

The equation $\mathbf{K}_n.\mathbf{R}_i = 2\pi n$ is the equation of a plane in real space, perpendicular to $\mathbf{K}_n$ and distant $2\pi n/|\mathbf{K}_n|$ from the origin, so that for different values of $n$ we have a set of parallel equidistant planes, all perpendicular to $\mathbf{K}_n$ and separated by $2\pi/|\mathbf{K}_n|$. The set may be labelled by the vector $\mathbf{K}_n$. The direct points $\mathbf{R}_i$ lie in these planes which determine the Bragg reflections. They are generally denoted by the Miller indices, such as (1,1,0), (3,2,1) ... which, in suitable units are just the components of the labelling vector $\mathbf{K}_n$. These concepts are very useful in the theory of X-ray diffraction by crystals.

The vectors $\mathbf{K}_n$ of the reciprocal lattice can in fact be obtained by observing the directions of strong X-ray reflections. Consider a monochromatic beam of X-rays, with propagation vector $\mathbf{k}$, incident on a crystal. Consider the reflected beam, of propagation vector $\mathbf{k}'$, due to the set of planes associated with

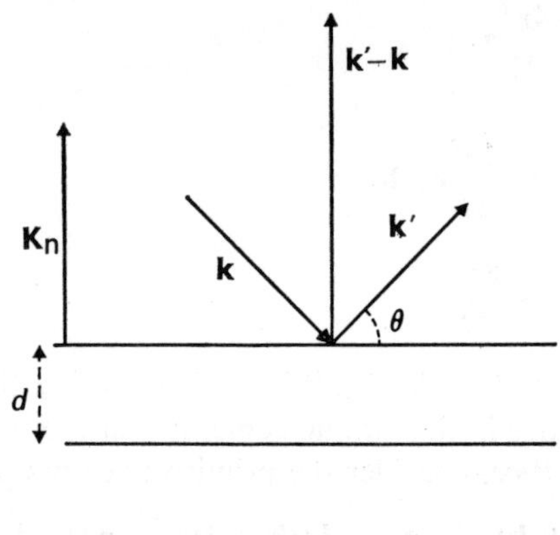


Fig. 2.13. Determination of reciprocal lattice vectors from X-ray reflections.

the vector $\mathbf{K}_n$. Of course for elastic reflection $|\mathbf{k}'| = |\mathbf{k}|$. From Fig. 2.13 we have

$$|\mathbf{k}' - \mathbf{k}| = 2k \sin \theta$$

Also, since $\mathbf{k}' - \mathbf{k}$ is parallel to $\mathbf{K}_n$,

$$\frac{(\mathbf{k}' - \mathbf{k})}{|\mathbf{k}' - \mathbf{k}|} = \frac{\mathbf{K}_n}{|\mathbf{K}_n|}$$

so that

$$\mathbf{k}' - \mathbf{k} = 2k \sin\theta \frac{\mathbf{K}_n}{|\mathbf{K}_n|}$$

But $|\mathbf{K}_n| = 2\pi/d$, and the Bragg condition gives

$$2d \sin\theta = p\lambda$$

where $p$ is an integer, $\lambda = 2\pi/k$, thus

$$\frac{1}{|\mathbf{K}_n|} = \frac{p\lambda}{4\pi \sin\theta} = \frac{p}{2k \sin\theta}$$

and

$$\mathbf{k}' = \mathbf{k} + p\mathbf{K}_n$$

a relation from which $\mathbf{K}_n$ may be obtained. The values of $p$ are obtained by observing several reflections from the same set of planes. For further details see, for instance, Lipson and Cochran, 1953.

## 2.8 Brillouin zones

Consider a state defined by a vector **k**, so that

$$\Psi_{\mathbf{k}}(\mathbf{r} + \mathbf{R}_i) = \exp(i\mathbf{k}.\mathbf{R}_i)\Psi_{\mathbf{k}}(\mathbf{r})$$

Because of (2.39) we can add to **k** any vector $\mathbf{K}_n$ of the reciprocal lattice, obtaining $\mathbf{k}' = \mathbf{k} + \mathbf{K}_n$, and we still can write

$$\Psi_{\mathbf{k}}(\mathbf{r} + \mathbf{R}_i) = \exp(i\mathbf{k}'.\mathbf{R}_i)\Psi_{\mathbf{k}}(\mathbf{r})$$

that is, the state $\Psi_{\mathbf{k}}$ satisfies Bloch's periodicity condition as if it had the wave vector $\mathbf{k}'$. The original label **k** is not unique: every state has an infinity of possible wave vectors, differing from one another by the fundamental vectors of the reciprocal lattice. This expresses again the fact that in the Bloch condition the phase change is defined only apart from multiples of $2\pi$.

The result may be expressed by saying that the one-electron wavefunctions of a perfect solid are periodic in reciprocal space, and then also the eigenvalues $E(\mathbf{k})$ must be periodic functions of **k**, with the periodicity of the reciprocal lattice. Similar considerations apply also when considering the frequency of the lattice vibrations as function of the propagation vector **q**. However for the lattice vibrations the wave is defined only at the lattice points so that all the vectors $\mathbf{k} + \mathbf{K}_n$ are exactly equivalent (Fig. 2.14). Thus for the vibration frequencies of a linear chain of equal atoms there is only one band. But the electronic wavefunctions are defined for all values of $x$, so that plane waves such as those of Fig. 2.14 belong to different bands and have different frequencies (energies). But the point is that the value of $\mathbf{K}_n$ to be assigned to any eigenvalue is arbitrary.

It is thus unnecessary to consider more than one of all the equivalent

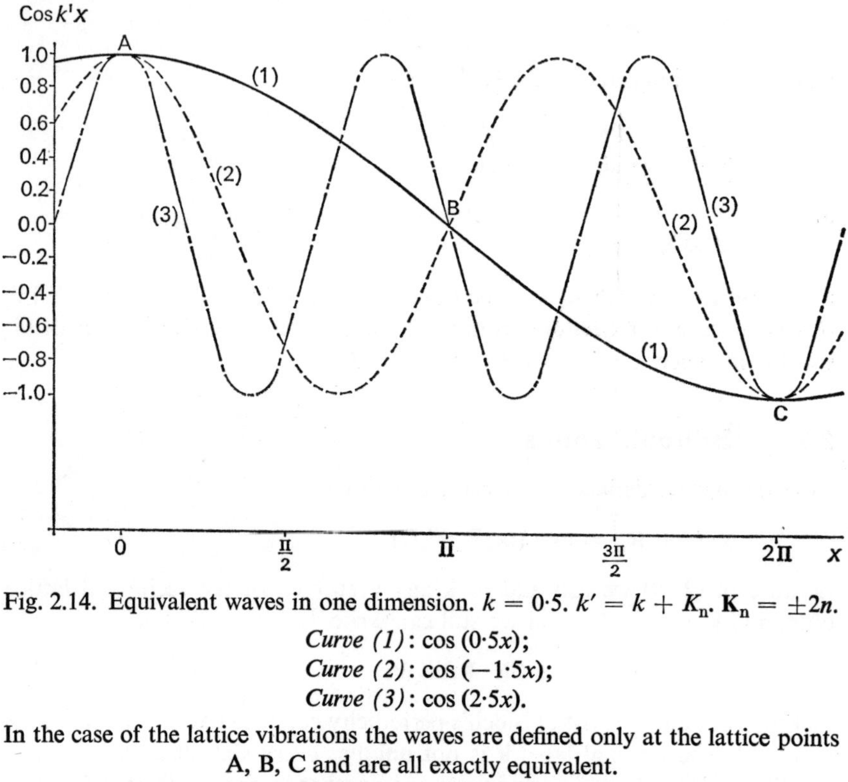


Fig. 2.14. Equivalent waves in one dimension. $k = 0{\cdot}5$. $k' = k + K_n$. $\mathbf{K}_n = \pm 2n$.

*Curve (1)*: $\cos(0{\cdot}5x)$;
*Curve (2)*: $\cos(-1{\cdot}5x)$;
*Curve (3)*: $\cos(2{\cdot}5x)$.

In the case of the lattice vibrations the waves are defined only at the lattice points A, B, C and are all exactly equivalent.

vectors $\mathbf{k} + \mathbf{K}_n$. The most convenient procedure is to reduce all the equivalent vectors to their smallest possible value that can be obtained by subtracting vectors of the reciprocal lattice. The end points of all the reduced vectors fill a region around the origin of reciprocal space, which is called the first, or central, Brillouin zone (Brillouin 1946). Since the other zones that can be constructed outside the first are not used often, we shall call the central zone just the Brillouin zone (BZ). It is the Wigner-Seitz cell of the reciprocal lattice. Its volume is thus the same as that of the primitive cell of the lattice, that is $8\pi^3/\Omega$. The second zone includes all the second smallest non-equivalent possible vectors, and so on.

In contrast to the direct lattice, where every unit cell is equivalent to every

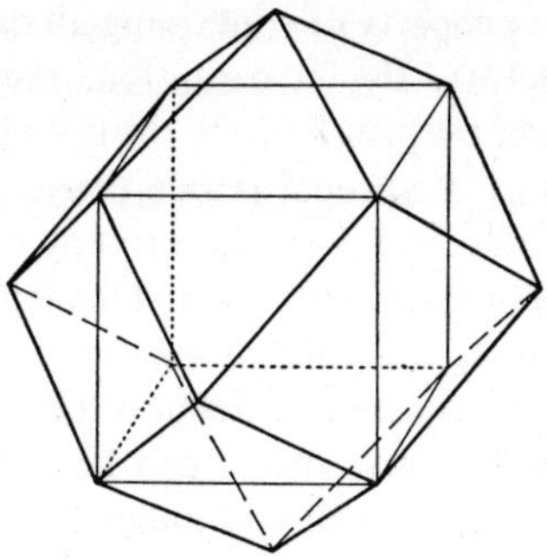

1st zone – central cube
2nd zone – between central cube and dodecahedron

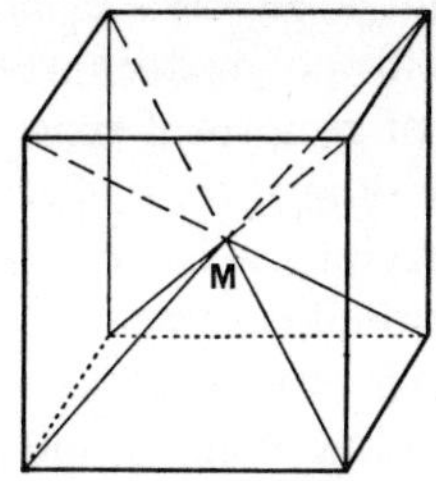


Reduction of 2nd zone

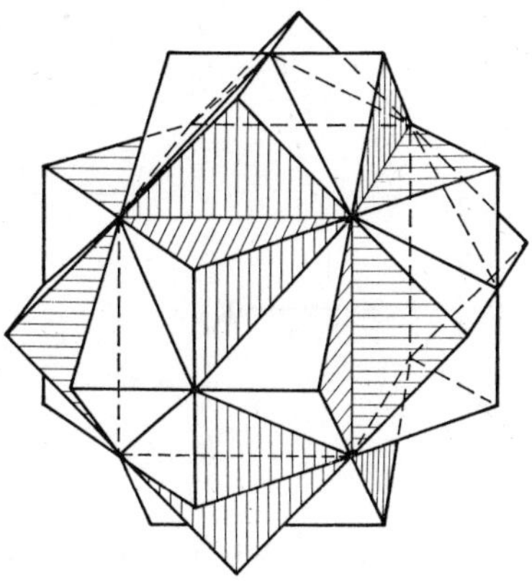

3rd zone – between dodecahedron and outer surface

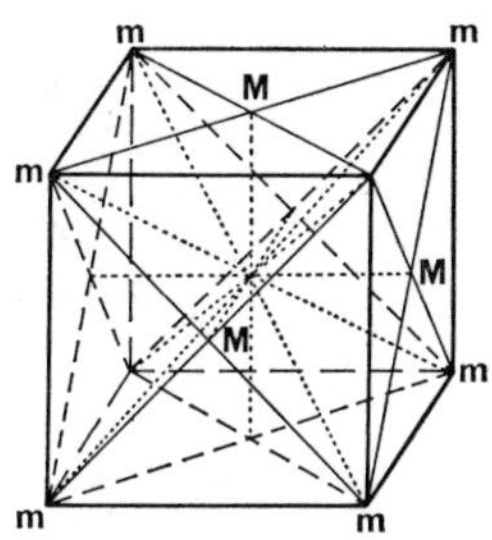


Reduction of 3rd zone

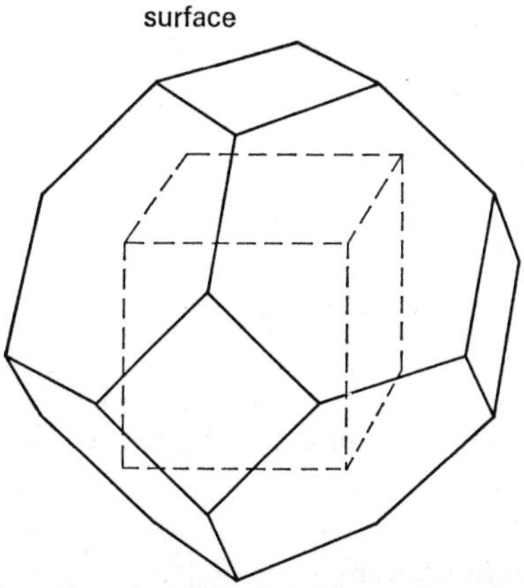

4th zone – between 3rd zone and truncated octahedron

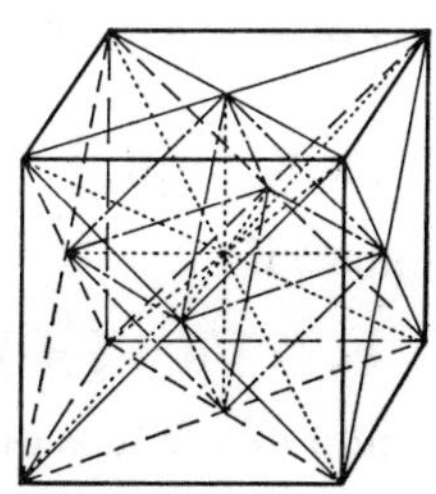

Reduction of 4th zone

Fig. 2.15. The first four Brillouin zones for the simple cubic lattice. (After Brillouin, 1946.)

other, the Brillouin zone has the unique property of containing all the smallest non-equivalent crystal momentum vectors. By its definition, the Brillouin zone is the polyhedron whose faces bisect perpendicularly the lines joining the origin of reciprocal space to the nearest reciprocal lattice points; it is that volume of reciprocal space which is closest to $\mathbf{k} = 0$ than to any other point of the reciprocal lattice. The other zones may be constructed in a similar way, though doubts sometimes arise in assigning certain regions to one or another zone. An unambiguous construction is indicated in Section 3.2. They have all the same volume and their various bits and pieces can be re-assembled to reproduce the central zone. The first four Brillouin zones for the simple cubic lattice are illustrated in Fig. 2.15.

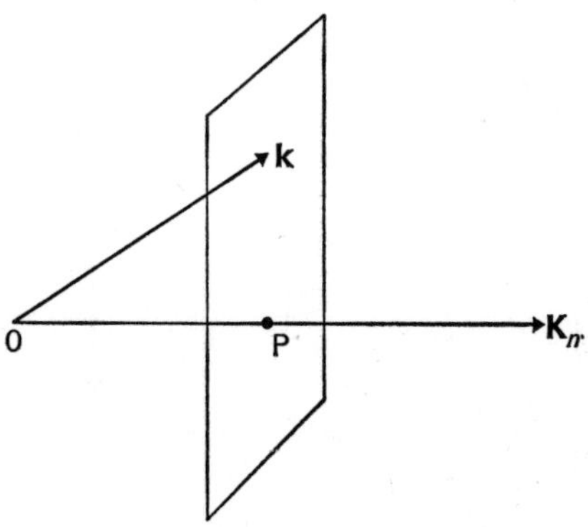


Fig. 2.16. Face of the Brillouin zone bisecting perpendicularly the reciprocal lattice vector $\mathbf{K}_n$.

The equation of the face of the Brillouin zone bisecting the reciprocal lattice vector $\mathbf{K}_n$ is obtained from (Fig. 2.16).

$$\overline{\mathrm{OP}} = \tfrac{1}{2}|\mathbf{K}_n| = \mathbf{k}.\mathbf{K}_n/|\mathbf{K}_n|$$

or

$$2\mathbf{k}.\mathbf{K}_n = |\mathbf{K}_n|^2 \tag{2.42}$$

For instance, if

$$\mathbf{K}_n = C(\mathbf{i} + \mathbf{j} + \mathbf{k})$$

the equation of the face of the Brillouin zone bisecting $\mathbf{K}_n$ perpendicularly is

$$k_x + k_y + k_z = 3C/2$$

Each face has a parallel face of equation $2\mathbf{k}.\mathbf{K}_n = -|\mathbf{K}_n|^2$.

Since the reciprocal lattice of a f.c.c.l. is b.c.c., the Brillouin zone for a f.c.c.l. is the truncated octahedron of Fig. 2.8 (Wigner-Seitz cell of the b.c.c.l.). The hexagonal faces bisect perpendicularly the lines joining the point $k = 0$ to its eight nearest neighbours along the [1,1,1] and equivalent directions (for crystallographic notations see, for instance, Phillips, 1963); the square faces bisect the lines joining $k = 0$ to its six second nearest neighbours along the [1,0,0] and equivalent directions. The Brillouin zone of the b.c.c.l. is a rhombododecahedron. The Brillouin zone of the hexagonal lattice is a hexagonal prism.

It should be stressed that the type of Brillouin zone is determined only by the type of Bravais lattice, not by the particular crystal structure. For non-cubic crystals however the actual shape of the Brillouin zone depends on the lattice parameters. There are thus just 14 types of Brillouin zone and, for instance, the Brillouin zones of Al, NaCl, Si, InSb are all the truncated octahedron characteristic of f.c.c. structures. However, the possible types of energy levels at symmetry points in the Brillouin zone, and the possible symmetries of the wavefunctions depend on the particular structure. In any case, it is necessary to determine energy values and wavefunctions only within, and on the surface of the first Brillouin zone. For any crystal, the same Brillouin zone is appropriate both for the vector **k** describing the electrons and for the vector **q** describing the lattice vibrations.

## 2.9 Zone schemes

The scheme in which all **k**-vectors are considered to lie in the Brillouin zone is called the reduced zone scheme. The energy is then a multivalued function of **k**. Sometimes it is more convenient to employ the extended zone scheme, bringing in the totality of **k**-space and assigning different bands to different Brillouin zones. The energy is then discontinuous at zone boundaries. When using the nearly free electrons (NFE) approximation (Section 5.3), it often happens that the energy is not discontinuous at some of these boundaries, for instance in the case of many ordered alloys. It is then useful to consider other zones, the Jones zones, on whose boundaries the energy is discontinuous also in the nearly free electrons approximation. Finally, having found the $E$–**k** relations, or the constant energy surfaces in the first Brillouin zone, it is sometimes useful to consider the same surfaces or relations in all cells of reciprocal space, as was done in Fig. 1.4. This is called repeated zone scheme.

A simple example will illustrate these possible schemes. Take a one-dimensional lattice with period $a$. The period of the reciprocal lattice is $2\pi/a$. The first Brillouin zone extends from $k = -\pi/a$ to $k = +\pi/a$. The second zone extends from $-2\pi/a$ to $-\pi/a$ and from $+\pi/a$ to $2\pi/a$, and so on. Suppose that the periodic potential is extremely weak, so that the $E(k)$ relation deviates little from $E = \hbar^2k^2/2m_0$. As we shall see (Section 5.3) the deviations occur at the zone boundaries where there are energy gaps and where the $E$–$\mathbf{k}$ curves must have horizontal tangents. Plotting the slightly modified parabola for all values of $k$ constitutes the extended zone scheme (Fig. 2.17a).

We now reduce all $\mathbf{k}$-vectors to the central zone by adding or subtracting an appropriate number of periods $2\pi/a$ of the reciprocal lattice. For instance,

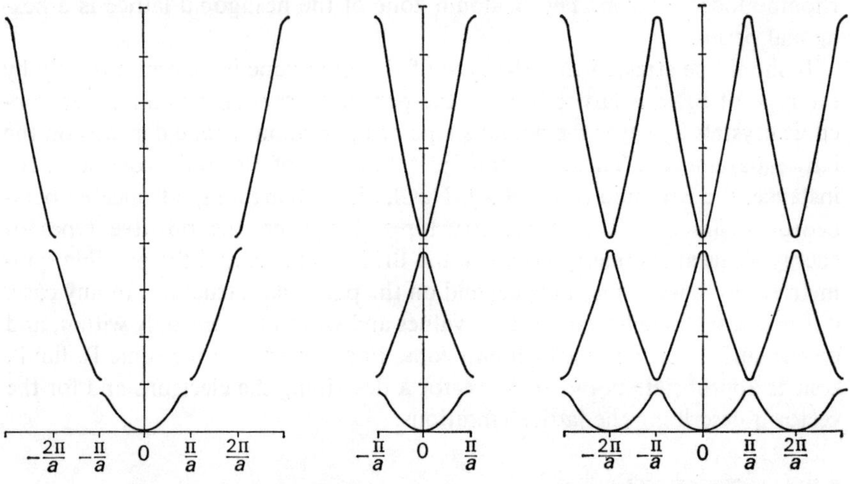


Fig. 2.17. Extended (a), reduced (b), and repeated (c) zone schemes in one dimension.

the point $+3\pi/2a$ is reduced to $3\pi/2a - 2\pi/a = -\pi/2a$. We obtain then the reduced zone scheme (Fig. 2.17b). $E$ is now a multivalued function of $k$. Note that every curve is a piece of the original parabola, displaced by $2\pi n/a$. Thus in this scheme the energy is given, for infinitely weak periodic potential, by

$$E_n(k) = \left(\frac{\hbar^2}{2m_0}\right)\left(k + \frac{2\pi n}{a}\right)^2 \quad n = 0, \pm 1, \pm 2 \ldots \tag{2.43}$$

and in general, in three dimensions

$$E_n(\mathbf{k}) = \frac{\hbar^2}{2m_0}|\mathbf{k} + \mathbf{K}_n|^2 \tag{2.44}$$

(free electron approximation, or empty lattice). For completely free electrons degeneracies occur at the zone boundaries; thus at $k = +\pi/a$, $E_0$ and $E_{-1}$ have the same energy. The degeneracy is removed by the periodic potential which introduces the energy gaps.

By repeating the curves of Fig. 2.17a throughout $k$-space, we obtain the repeated zone scheme (Fig. 2.17c), which brings out the periodicity of $E$ with

$k$. Which scheme to use in any particular problem is purely a matter of convenience. For the purpose of calculating band structures the reduced zone scheme is the most useful. The $E$–$\mathbf{k}$ curves for the numerical example of Section 2.3 are similar to those of Fig. 2.17.

A two-dimensional example making use of constant energy curves is illustrated in Fig. 2.19. The lattice here is a square lattice and the potential has been taken as

$$V = 0{\cdot}2\,(\cos 2x + \cos 2y) \tag{2.45}$$

Schrödinger's equation is then separable and the separated equations can be treated by perturbation theory, by considering the potential as weak (NFE approximation, see Section 5.3, where the same potential is considered).

Fig. 2.18 illustrates the first two Brillouin zones and how the second zone

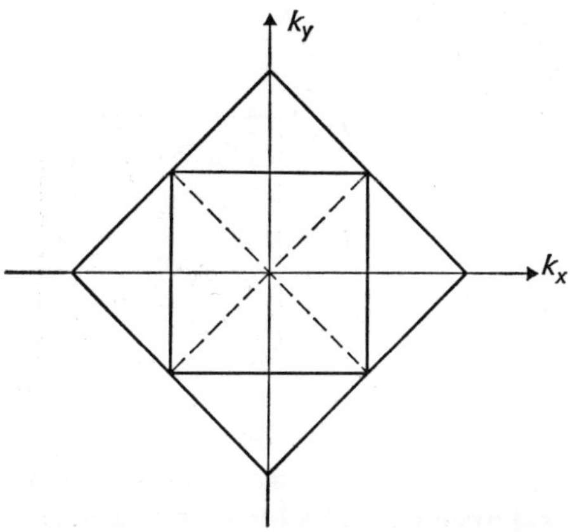


Fig. 2.18. The first two Brillouin zones for a square lattice.

can be reduced to the first by translating the various pieces. Fig. 2.19a shows the constant energy curves in the extended zone scheme. The lowest energy level is assigned to the first zone, the second lowest to the second zone, etc. $E(\mathbf{k})$ is discontinuous across the boundary of the two zones. In the nearly free electron approximation the discontinuity is constant all over the boundary, and equal to 0·2. Fig. 2.19b shows the energy curves of the second zone in the reduced zone scheme. Each zone can be repeated, and the repeated scheme for the first zone is shown in Fig. 2.19c, where $E(\mathbf{k})$ is periodic and continuous across the boundaries of the cells for each zone.

In the simplest cases the energies of a zone form a band. In the extended zone scheme each state is assigned a different $\mathbf{k}$. This scheme is useful for simple metals. The inner core electrons are not generally considered. For instance, in Na, the state $\mathbf{k} = 0$ is assigned to the bottom of the band deriving from the $3s$ level. This band fills the first Brillouin zone and the band derived from the triply degenerate $3p$ level fills the second, third and fourth Brillouin

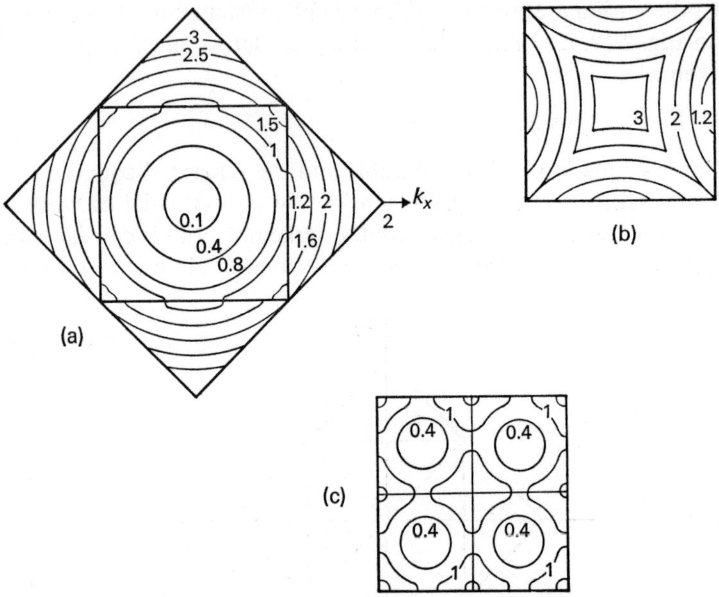


Fig. 2.19. Constant energy curves for square lattice with potential $V = 0{\cdot}2(\cos 2x + \cos 2y)$.

(a) Extended zone scheme: first two zones.
(b) Reduced zone scheme: second zone.
(c) Repeated zone scheme: first zone.

zones, but the scheme breaks down for higher bands (For a discussion see Slater, 1934).

## 2.10 Band index

In the reduced zone scheme $E$ is a multivalued function of $\mathbf{k}$ and the successive values for any given $\mathbf{k}$ correspond first to the inner core electrons; then, in a semiconductor, one value, or more than one if degeneracy occurs, corresponds to the valence band, and the next value, or values, to the conduction band; higher values correspond to higher unoccupied bands. Thus a fourth quantum number, generally called the band index, $n$, denoting the particular band, is necessary to identify completely any state. Bands are usually given an index increasing in order of increasing energy, but difficulties arise because the prescription breaks down at degeneracy points. To illustrate: if along a direction in **k**-space we have a point of degeneracy, A, (Fig. 2.20), our rule interchanges the indices 1,2 of the two curves in passing through A. It follows that $E_i$ is not a differentiable function of **k** at A. Note however that the direc-

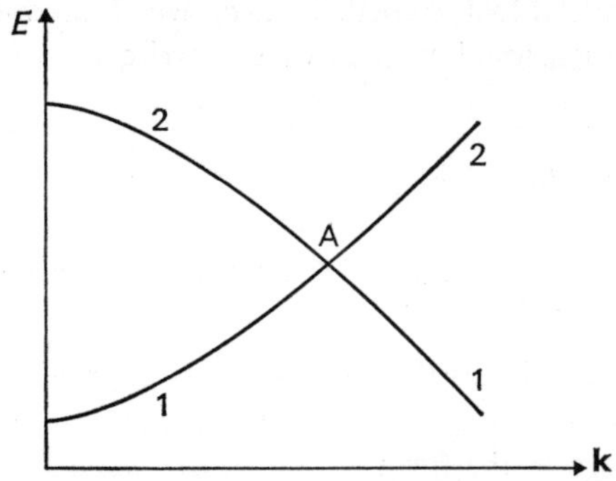


Fig. 2.20. Crossing of two energy bands. The bank index is exchanged.

tion of **k** must be along a line of symmetry for a degeneracy to occur, otherwise, if there is no symmetry, we shall see that all bands are of the same type: in the language of group theory (Section 6.1), they belong to the same irreducible representation – the identity – of the group of **k**; in these conditions they cannot cross, as can be seen by considering that the various energies of the same type can be found as the roots of a determinantal equation, as explained in Section 2.4, and it is well known that, if the coefficients of the equation depend on a parameter, the roots never cross when the parameter varies. We see that the points of degeneracy form at most lines in **k**-space.

When the electrons are tightly bound, the band index may be taken to coincide with the spectroscopic notation for the atomic levels, e.g. 1*s*, 2*s*, 2*p* ... When the electrons are almost free, formula (2.44) indicates that the levels are properly denoted by vectors $\mathbf{K}_n$ of the reciprocal lattice, though, in view of the possibility of crossing, $\mathbf{K}_n$ may, in the same band, be different in different regions of **k**-space. Fig. 2.17 has examples of this.

## 2.11 Aims of band theory

The main problem of band theory is then the determination of the $E_i$ values of interest for any **k** inside or on the surface of the BZ, and particularly near those points at which $E$ is an extremum. In semiconductors, electrons and holes are located at these points and the nature of the dispersion relations at these points determines many of the properties of the substance. Near an extremum of energy $E^0$ at a point $\mathbf{k}^0$ a Taylor expansion gives

$$E(\mathbf{k}) = E(\mathbf{k}^0) + \sum_{j,i=x,y,z} c_{ij}(k_i - k_i^0)(k_j - k_j^0) + \ldots \tag{2.46}$$

Higher terms can often be neglected, and then the dispersion relation near these points is expressed by the coefficients $c_{ij}$ which are the nine components of a tensor. For dimensional reasons we can write

$$c_{ij} = \frac{\hbar^2}{2}\left(\frac{1}{m^*}\right)_{ij} \tag{2.47}$$

and the quantities $\left(\frac{1}{m^*}\right)_{ij}$ are considered as the components of an 'inverse mass tensor'. This constitutes the effective mass approximation (Section 4.5). By choosing the coordinate axes along the principal axes of the tensor, (2.46) can always be written in the form

$$E(\mathbf{k}) = E(\mathbf{k}^0) + \sum_{i=1}^{3} \frac{\hbar^2}{2m_i}(k_i - k_i^0)^2 \tag{2.48}$$

where the $m_i$ are called the principal masses (Section 4.4).

In metals at 0 K the occupied states fill some bounded region of **k**-space: all states whose energy is less than a certain energy, the Fermi energy, or Fermi level, $E_F$, are full. All other states are empty. The surface bounding the occupied states is called the Fermi surface (FS). The existence of a Fermi surface across which the occupation number of the one-electron states is discontinuous is an almost exact result of the quantum theory of many-fermion systems. The determination of the Fermi surface of metals is another problem of band theory. In principle it requires a calculation of all the energy levels and a determination of the number of states at every energy below $E_F$. Except for the alkali metals, the Fermi surface is not a sphere, but a very complicated surface. We shall see that fortunately it is possible to give a good qualitative account of the Fermi surface of many metals by using the free electron approximation.

In one dimension the energy bands are always separated by forbidden energy gaps (with unimportant exceptions). In three dimensions, overlapping occurs very commonly, that is $E_n(\mathbf{k}_1)$ may be larger than $E_{n+1}(\mathbf{k}_2)$ at many points $\mathbf{k}_1$, $\mathbf{k}_2$. Also, if the crystal contains screw axes or glide planes, two bands may meet at points or over an area of the boundary of the Brillouin zone.

If a face of the Brillouin zone is parallel to a reflection plane, as are faces

$a$, $b$ in Fig. 2.21 then $\partial E/\partial n$, where $\mathbf{n}$ is the outer normal to the face must vanish at all points of this surface. In fact $E(A) = E(B)$ because of reflection symmetry; hence

$$\left(\frac{\partial E}{\partial n}\right)_A = +\left(\frac{\partial E}{\partial n}\right)_B$$

But $A$, $B$ are equivalent points, since they differ by a reciprocal lattice vector, hence

$$\left(\frac{\partial E}{\partial n}\right)_A = -\left(\frac{\partial E}{\partial n}\right)_B$$

Thus $(\partial E/\partial n)$ must vanish at all points of $a$, $b$. Considering, for example, the truncated octahedron of Fig. 2.8, $\partial E/\partial n$ vanishes on the square faces, but not on the hexagonal faces which are not parallel to a reflection plane. It is easy to

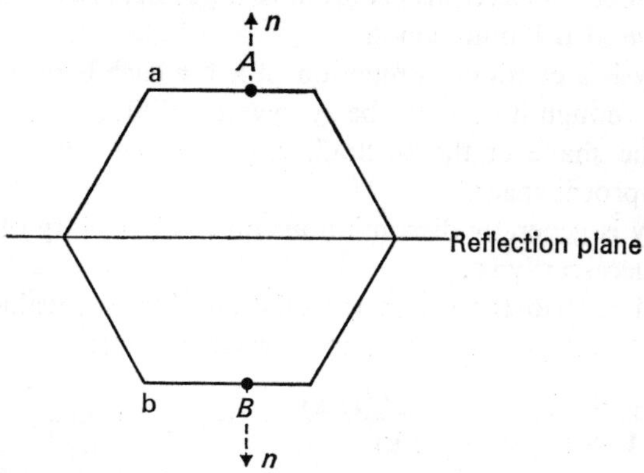


Fig. 2.21. Section of Brillouin zone with faces a, b parallel to a reflection plane.

show, however, that it must vanish on lines of symmetry on these faces. An equivalent way of stating the theorem is to say that the constant energy surfaces must cut perpendicularly faces such as $a$, $b$ (Cfr. Fig. 2.19).

If along a direction $\mathbf{n}$ perpendicular to a face of the Brillouin zone two energy curves meet at the boundary and are thus degenerate (this occurs in crystals with screw axes), they must satisfy the relation

$$\frac{\partial E_1}{\partial n} + \frac{\partial E_2}{\partial n} = 0$$

As we saw in Section 2.1, each non-degenerate band can accommodate two electrons with opposite spin for each primitive cell of the crystal, that is for each allowed $\mathbf{k}$-vector: there are exactly as many allowed wave vectors in the Brillouin zone as there are unit cells in the block of the crystal. Now, if $\Omega$ is the volume of the unit cell of the crystal, we know that the volume of the Brillouin zone is $8\pi^3/\Omega$. If there are $N$ cells in the volume $V$, $\Omega = V/N$. The volume of **k**-space per allowed **k**-vector is thus $8\pi^3/V$, the extension to three dimensions of what we found in the one-dimensional treatment of Section 2.1. Thus there are $V/8\pi^3$ allowed **k**-vectors per unit volume of reciprocal space and, introducing the factor 2 due to spin, there are $V/4\pi^3$ states per unit volume of reciprocal space. This density of states is thus constant throughout **k**-space; it increases when the size of the crystal increases and is independent of the zone structure.

Summarising, the Brillouin zone has the following main properties:

1. It contains two electron states per band for each primitive Bravais cell of the crystal. If the crystal has a simple Bravais lattice, so that each unit cell contains only one atom, a Brillouin zone accommodates two electrons per atom. If the number of electrons per atom is larger than two, the electrons are spread over several Brillouin zones.
2. The energy is a continuous function of **k** for each band throughout the Brillouin zone, though it may not be everywhere differentiable.
3. Cells of the shape of the Brillouin zone can be fitted together to fill completely reciprocal space.
4. The energy is generally discontinuous on the boundary of the Brillouin zone, but not necessarily so.

We recapitulate also the symmetry relations for eigenvalues and eigenfunctions of Schrödinger's equation in a perfect crystal:

$$\begin{aligned} &\Psi_n(\mathbf{r} + \mathbf{R}_i; \mathbf{k}) = e^{i\mathbf{k}.\mathbf{R}_i}\,\Psi_n(\mathbf{r}; \mathbf{k}) \\ &\Psi_n(\mathbf{r}; \mathbf{k} + \mathbf{K}_i) = \Psi_n(\mathbf{r}; \mathbf{k}) \\ &E_n(\mathbf{k} + \mathbf{K}_i) = E_n(\mathbf{k}) \\ &\Psi_n(\mathbf{r}; -\mathbf{k}) = \Psi_n^*(\mathbf{r}; \mathbf{k}) \\ &E_n(-\mathbf{k}) = E_n(\mathbf{k}) \end{aligned} \tag{2.49}$$

## 2.12 Density of states

We have seen that, for a unit volume of the crystal, the number of states in volume $d\mathbf{k}$ of reciprocal space is

$$dn_s = \frac{1}{4\pi^3}\,d\mathbf{k} \tag{2.50}$$

The number of states per unit energy interval, $dn_s/dE$, is called the density of states. Knowledge of the density of states is necessary for the interpretation of many phenomena. In particular the density of states at the Fermi level is of great importance in determining the properties of metals, for instance the electronic specific heat.

Since the density of states in **k**-space has the constant value $1/4\pi^3$, the number of states in a volume $A$ of **k**-space is

$$n_s = \frac{1}{4\pi^3}\int_A d\mathbf{k} \quad \text{(per unit volume of the crystal)} \tag{2.51}$$

Choose $d\mathbf{k}$ to be bounded on two faces by surfaces of constant energy, then

$$d\mathbf{k} = \frac{dS\,dE}{|\text{grad}_\mathbf{k}\, E|}$$

where $dS$ is an element of area of the constant energy surface. Then the density of states is

$$\frac{dn_s}{dE} = \frac{1}{4\pi^3}\int_A \frac{dS}{|\text{grad}_\mathbf{k}\, E|} \tag{2.52}$$

the integral extending over the constant energy surface $E(\mathbf{k}) = E$ in **k**-space. We shall see that $\text{grad}_\mathbf{k}\, E$ is $\hbar$ times the velocity of the electron in state **k**, as is plausible since $\hbar^{-1}\,\text{grad}_\mathbf{k}\, E$ is the group velocity of the associated waves.

As an example, suppose that

$$E = E^0 + \frac{\hbar^2 k^2}{2m^*}$$

where $m^*$ is an effective mass. Then

$$\text{grad}_\mathbf{k}\, E = \frac{\hbar^2 \mathbf{k}}{m^*}$$

so that

$$\frac{dn_s}{dE} = \frac{1}{4\pi^3}\frac{m^*}{\hbar^2}\int_A \frac{dS}{k}$$

The constant energy surfaces are spheres, thus

$$\frac{dn_s}{dE} = \frac{m^*}{4\pi^3\hbar^2}4\pi k = \frac{m^* k}{\pi^2\hbar^2} \tag{2.53}$$

Since

$$(E - E^0)^{\frac{1}{2}} = \frac{\hbar k}{(2m^*)^{\frac{1}{2}}}$$

we can also write

$$dn_s = \frac{(2m^{*3})^{\frac{1}{2}}}{\pi^2\hbar^3}(E - E^0)^{\frac{1}{2}}\, dE \tag{2.54}$$

This formula applies to semiconductors within the effective mass approximation (Section 4.5). The variation of $dn_s/dE$ with $E$ is illustrated in Fig. 2.22 for $m^* = m_0$.

It is evident that $dn_s/dE$ may have singularities if the integration includes points at which $\text{grad}_\mathbf{k}\, E = 0$. These points are called critical points, and are

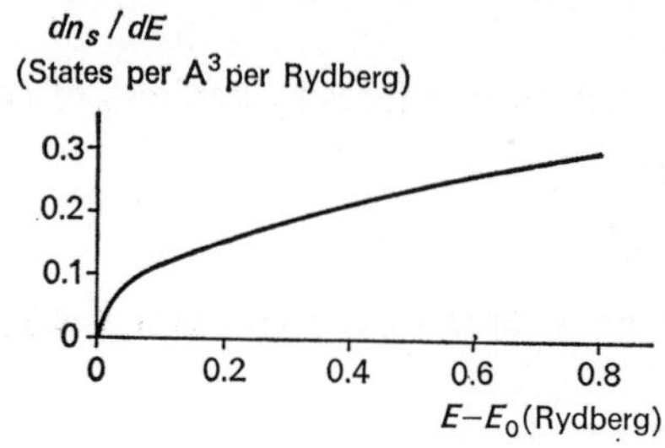


Fig. 2.22. Density of states in isotropic parabolic band with $m = m_0$.

always present. Where $\text{grad}_\mathbf{k}\, E$ vanishes, $E(\mathbf{k})$ has the form (2.48); there are four possible types of critical points, depending on the values of the $m_i$:

$P_1$: $m_1$, $m_2$, $m_3$ all negative (a maximum);
$P_2$: $m_1$, $m_2$ negative, $m_3$ positive (saddle point of first kind);
$P_3$: $m_1$ negative, $m_2$, $m_3$ positive (saddle point of second kind);
$P_4$: $m_1$, $m_2$, $m_3$ all positive (a minimum).

It can be proved that, in three dimensions, the minimum set of critical points in any band is one maximum, one minimum and six saddle points, three of each kind, and that the density of states is continuous at a critical

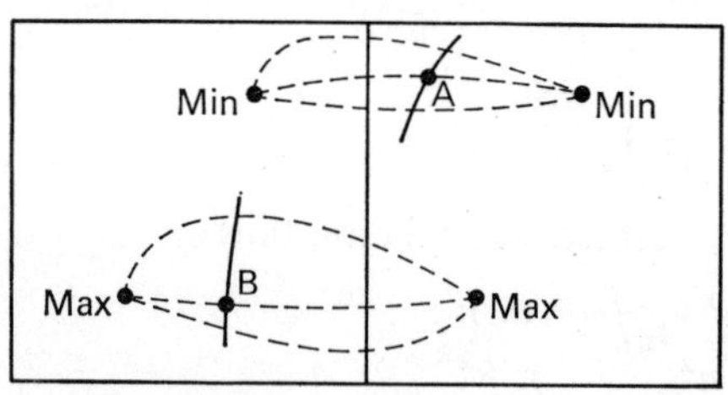


Fig. 2.23. Critical points for square lattice.

point, but its first derivatives with respect to $k_x$, $k_y$, $k_z$ have discontinuities. In two dimensions there are at least one maximum, one minimum and two saddle points. An example is illustrated in Fig. 2.23. On the curves connecting two absolute minima in adjacent cells there are relative maxima. Join all these maxima (solid curve): the lowest relative maximum, A, on this curve is a saddle point. Similarly, the highest relative minimum, B, on the curves joining two absolute maxima is another saddle point. In the example of Fig. 2.23, for symmetry reasons, there are in each cell eight maxima, eight minima and sixteen saddle points. However, critical points generally occur at symmetry points and then there are not so many of them.

## 2.13 Non-periodic structures

Although energy bands are characteristic of periodic structures, we have seen that bands exist in a disordered material, being broadened atomic levels. For these bands it is still possible to consider a dispersion relation, but the wavefunctions have not in general the Bloch form. However it has been shown (see for instance Phariseau and Ziman, 1963; Mott, 1967; Edwards 1967) that a Bloch condition can be introduced for a disordered structure, such as a liquid metal, provided it is homogeneous. As a consequence of disorder, the vector **k** becomes complex, $\mathbf{k} = \mathbf{k}_1 + i\mathbf{k}_2$, which corresponds to a travelling wave decaying like $e^{-\mathbf{k}_2 r}$ in the direction of propagation. The reciprocal of the modulus of $\mathbf{k}_2$ is thus essentially the mean free path. The disorder scatters the electronic waves into states with other **k**-values allowed by conservation of energy.

It is found experimentally that the loss of long-range order does not have a pronounced effect on the electric and other properties, and thus presumably on the band structure, provided the short-range order is preserved, that is, provided the coordination corresponding to the solid phase is preserved in the liquid phase. Thus, when a metal with a close-packed structure melts, the coordination remains more or less the same, and few discontinuities occur. On the other hand, fusion has a marked effect on substances with directed bonds that must necessarily be broken at the melting point. For instance, in germanium the density increases at melting point and the open diamond structure changes into one approaching that of a close-packed metal. The band structure can then be expected to change drastically. The conductivity increases due to the reduction, or disappearance of the energy gap: most semiconductors become metallic on melting.

Also in liquids and other disordered structures, some bands, deriving from fully occupied atomic levels, may be completely full, so that the possibility of conduction by holes remains.

There have been several interesting approaches to the problem of calculating the electronic energy levels and the density of states in a non-periodic structure (see for instance Mott, 1967), but no complete solution has been obtained. It appears that, within a given range of energies, states are either localised or non-localised. In the former case, an electron originally in a certain small region of the crystal does not diffuse away. The energy ranges in which electrons are localised or non-localised are separated by certain critical energies at which the mean free path $L$ approaches the electron wavelength $\lambda$; in fact, values of $L$ less than $\lambda/2\pi$ cannot occur.

There is much experimental evidence that at the Fermi surface of liquid metals or disordered alloys states are not localised; but localised states exist in 'tails' on either side of each band in amorphous materials. In semiconductors there are bound states at impurity centres (Section 4.2), but, if the overlap between the wavefunctions of electrons at different centres is large, an 'impurity band' may arise with essentially metallic conduction. If the semiconductor contains both donors and acceptors, there may be 'hopping' motion between one centre and the other. This type of conduction involves an activation energy.

In the case of an amorphous semiconductor, band conduction takes place inside bands, while hopping motion, with a much reduced mobility, occurs in the tails extending into what would be the energy gap in the crystal. For these substances it is then more appropriate to speak of a mobility gap, rather than of an energy gap.

# 3

# Formalism of band theory

## 3.1 Fourier expansions

We have already remarked that any function with the periodicity of the lattice can be expanded in a Fourier series in terms of the vectors $\mathbf{K}_n$ of the reciprocal lattice. Thus we considered the expansion (2.40) of the potential $V(r) = \Sigma_{\mathbf{K}_n} V(\mathbf{K}_n)\, e^{i\mathbf{K}_n \cdot \mathbf{r}}$. Each Fourier coefficient is associated with a point in the reciprocal lattice. If there is more than one atom in the unit cell, some of the $V(\mathbf{K}_n)$ may vanish, or not be independent. In general the coefficients are the same for $\mathbf{K}_n$-vectors of the same length, for example, for the eight vectors $(\pm K, \pm K, \pm K)$. The $V(\mathbf{K}_n)$ are real if the crystal has a centre of inversion at $k = 0$; otherwise,

$$V(-\mathbf{K}_n) = V^*(\mathbf{K}_n) \tag{3.1}$$

The coefficients $V(\mathbf{K}_n)$ are found by the usual integral expression, but here we have the advantage that the periodic functions $e^{i\mathbf{K}_n \cdot \mathbf{r}}$ form a complete orthogonal system within the unit cell, which may be taken either as the primitive cell of the Bravais lattice, or as the Wigner-Seitz cell, or as any other equivalent cell. To prove the theorem, we use cells defined by the vectors $\mathbf{a}_1, \mathbf{a}_2, \mathbf{a}_3$ in the direct lattice and by $\mathbf{b}_1, \mathbf{b}_2, \mathbf{b}_3$ in the reciprocal lattice, and we write

$$\mathbf{r} = r_1\mathbf{a}_1 + r_2\mathbf{a}_2 + r_3\mathbf{a}_3$$

that is

$$x = r_1 a_{1x} + r_2 a_{2x} + r_3 a_{3x}, \qquad \text{etc} \tag{3.2}$$

where $r_1, r_2, r_3$ can take any value, not only integral values;

$$\mathbf{K}_n = n_1\mathbf{b}_1 + n_2\mathbf{b}_2 + n_3\mathbf{b}_3 \quad (n_1, n_2, n_3 \text{ integers}) \tag{3.3}$$

Multiply (2.40) by $e^{-i\mathbf{K}_m \cdot \mathbf{r}}$

$$\text{where } \mathbf{K}_m = m_1\mathbf{b}_1 + m_2\mathbf{b}_2 + m_3\mathbf{b}_3 \tag{3.4}$$

($m_1$, $m_2$, $m_3$ integers) and integrate over the unit cell

$$\int_{\text{unit cell}} V(\mathbf{r})e^{-i\mathbf{K}_m.\mathbf{r}}\, d\mathbf{r} =$$

$$\sum_{n_1,n_2,n_3} V(\mathbf{K}_n) \int_{\text{unit cell}} \exp\{i[(n_1 - m_1)\mathbf{b}_1 + (n_2 - m_2)\mathbf{b}_2 + (n_3 - m_3)\mathbf{b}_3].\mathbf{r}\}\, dx\, dy\, dz \quad (3.5)$$

We now consider $r_1$, $r_2$, $r_3$ as variables instead of $x$, $y$, $z$: when each of these quantities ranges from 0 to 1, the unit cell is exactly covered. Thus the integral in (3.5) becomes

$$\int_0^1 \exp\{i[(n_1 - m_1)\ {}_1 + (n_2 - m_2)\mathbf{b}_2 + (n_3 - m_3)\mathbf{b}_3] \,.\, (r_1\mathbf{a}_1 + r_2\mathbf{a}_2 + r_3\mathbf{a}_3)\}J\, dr_1\, dr_2\, dr_3$$

where the Jacobian $J$ is equal to

$$J = \begin{vmatrix} \dfrac{\partial x}{\partial r_1} & \dfrac{\partial x}{\partial r_2} & \dfrac{\partial x}{\partial r_3} \\ \dfrac{\partial y}{\partial r_1} & \dfrac{\partial y}{\partial r_2} & \dfrac{\partial y}{\partial r_3} \\ \dfrac{\partial z}{\partial r_1} & \dfrac{\partial z}{\partial r_2} & \dfrac{\partial z}{\partial r_3} \end{vmatrix} = \begin{vmatrix} a_{1x} & a_{2x} & a_{3x} \\ a_{1y} & a_{2y} & a_{3y} \\ a_{1z} & a_{2z} & a_{3z} \end{vmatrix} = \Omega$$

the volume of the unit cell. Now using (2.36) the integral becomes

$$\Omega \int_0^1 e^{2\pi i(n_1-m_1)r_1}\, dr_1 \int_0^1 e^{2\pi i(n_2-m_2)r_2}\, dr_2 \int_0^1 e^{2\pi i(n_3-m_3)r_3}\, dr_3 = \Omega\delta(\mathbf{K}_m - \mathbf{K}_n).$$

Thus

$$V(\mathbf{K}_m) = \frac{1}{\Omega}\int_{\text{one cell}} V(\mathbf{r})\, e^{-i\mathbf{K}_m\cdot\mathbf{r}}\, d\mathbf{r} \quad (3.6)$$

## 3.2 Momentum eigenfunctions

The other periodic function we have considered is $u_{\mathbf{k}}(\mathbf{r})$, the periodic factor of the Bloch function. Omitting the band index, this can then be expanded as

$$u_{\mathbf{k}}(\mathbf{r}) = \Sigma_{\mathbf{K}_n}\, v_{\mathbf{k}}(\mathbf{K}_n)\, e^{i\mathbf{K}_n\cdot\mathbf{r}} \quad (3.7)$$

with

$$v_{\mathbf{k}}(\mathbf{K}_n) = \frac{1}{\Omega}\int_{\text{one cell}} u_{\mathbf{k}}(\mathbf{r})\, e^{-i\mathbf{K}_n\cdot\mathbf{r}}\, d\mathbf{r} \quad (3.8)$$

The $v_{\mathbf{k}}(\mathbf{K}_n)$ are called momentum eigenfunctions. For each **k** and each band there is a set of numbers $v_{\mathbf{k}}(\mathbf{K}_n)$, one for each vector of the reciprocal lattice. For reasons that will become clear later, they are conveniently denoted by $v(\mathbf{k} - \mathbf{K}_n)$. Multiplying now (3.7) by $e^{i\mathbf{k}.\mathbf{r}}$, we have

$$\Psi_{\mathbf{k}}(\mathbf{r}) = \Sigma_{\mathbf{K}_n} v(\mathbf{k} - \mathbf{K}_n)\, e^{i(\mathbf{k}+\mathbf{K}_n).\mathbf{r}} \tag{3.9}$$

and since also the waves $\exp[i(\mathbf{k} + \mathbf{K}_n).\mathbf{r}]$ form an orthogonal set in the unit cell, (3.9) can be inverted, to give

$$v(\mathbf{k} - \mathbf{K}_n) = \frac{1}{\Omega}\int_{\text{unit cell}} \Psi_{\mathbf{k}}(\mathbf{r})\, e^{-i(\mathbf{k}+\mathbf{K}_n).\mathbf{r}}\, d\mathbf{r} \tag{3.10}$$

If $\Psi$ is normalised in the unit cell, we have

$$\begin{aligned} 1 &= \int_{\text{unit cell}} \Psi^*_{\mathbf{k}}(\mathbf{r})\Psi_{\mathbf{k}}(\mathbf{r})\, d\mathbf{r} \\ &= \Sigma_{\mathbf{K}_m,\mathbf{K}_n} v^*(\mathbf{k} - \mathbf{K}_n)v(\mathbf{k} - \mathbf{K}_m)\int_{\text{unit cell}} e^{i(\mathbf{K}_m-\mathbf{K}_n).\mathbf{r}}\, d\mathbf{r} \\ &= \Omega\, \Sigma_{\mathbf{K}_m,\mathbf{K}_n} v^*(\mathbf{k} - \mathbf{K}_n)v(\mathbf{k} - \mathbf{K}_m)\delta(\mathbf{K}_n - \mathbf{K}_m) \\ &= \Omega\, \Sigma_{\mathbf{K}_n} v^*(\mathbf{k} - \mathbf{K}_n)v(\mathbf{k} - \mathbf{K}_n) \end{aligned}$$

Thus the orthonormalisation condition for the momentum eigenfunction, when $\Psi$ is normalised in the unit cell, is

$$\Sigma_{\mathbf{K}_m,\mathbf{K}_n} v^*(\mathbf{k} - \mathbf{K}_n)v(\mathbf{k} - \mathbf{K}_m) = \frac{1}{\Omega}\delta(\mathbf{K}_m - \mathbf{K}_n) \tag{3.11}$$

There is a similar condition for $v$'s belonging to different bands $i, j$:

$$\Sigma_{\mathbf{K}_n} v_i^*(\mathbf{k} - \mathbf{K}_n)v_j(\mathbf{k} - \mathbf{K}_n) = \frac{1}{\Omega}\delta_{ij} \tag{3.12}$$

With the above normalisation condition, the square of the modulus of $v(\mathbf{k} - \mathbf{K}_n)$, multiplied by $\Omega$, gives the probability that the electron in state **k** has momentum $\hbar(\mathbf{k} + \mathbf{K}_n)$. This follows from the general postulates of quantum mechanics, since the plane waves are the eigenfunctions of the momentum operator. However, the bands must be well separated, so that there are no non-diagonal matrix elements of momentum with respect to band index. The average momentum is

$$\bar{p}(\mathbf{k}) = \Sigma_{\mathbf{K}_n} v^*(\mathbf{k} - \mathbf{K}_n)\hbar(\mathbf{k} + \mathbf{K}_n)v(\mathbf{k} - \mathbf{K}_n) \tag{3.13}$$

and similarly for any function of momentum. Only a discrete set of values of momentum exist for an electron in state **k**†.

† The expression for the average momentum in terms of the Bloch functions is given by the diagonal elements of (3.38).

The average velocity is

$$\overline{\mathbf{w}}(\mathbf{k}) = \frac{1}{m_0}\,\overline{\mathbf{p}}(\mathbf{k})$$

but (equation 4.18)

$$\overline{\mathbf{w}}(\mathbf{k}) = \frac{1}{\hbar}\,\mathrm{grad}_{\mathbf{k}}\,E,$$

hence we have an expression for $\mathrm{grad}_{\mathbf{k}}\,E$ in terms of the momentum eigenfunctions:

$$\mathrm{grad}_{\mathbf{k}}\,E = \frac{\hbar^2}{m_0}\,\Sigma_{\mathbf{K}_n}\,v^*(\mathbf{k}-\mathbf{K}_n)(\mathbf{k}+\mathbf{K}_n)v(\mathbf{k}-\mathbf{K}_n)$$

If we let **k** take all values inside the Brillouin zone, all the coefficients vary, and are in fact continuous functions of **k**. But, because of the periodicity in **k**-space, they all are in fact the same function of **k**, only displaced one with

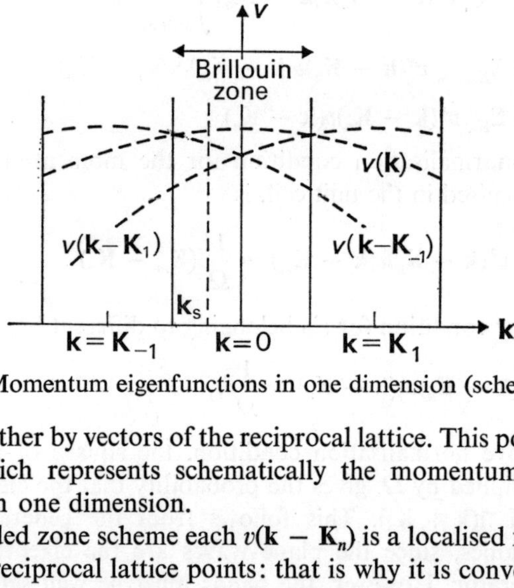


Fig. 3.1. Momentum eigenfunctions in one dimension (schematic).

respect to the other by vectors of the reciprocal lattice. This point is illustrated in Fig. 3.1 which represents schematically the momentum eigenfunctions for one band in one dimension.

In the extended zone scheme each $v(\mathbf{k}-\mathbf{K}_n)$ is a localised function centred on one of the reciprocal lattice points: that is why it is convenient to denote them as $v(\mathbf{k}-\mathbf{K}_n)$. They are all identical, so that we can speak of one function $v(\mathbf{k})$. This function determines the probability distribution of momentum for the whole band. If all the $\Psi_{\mathbf{k}}(\mathbf{r})$ are normalised in the unit cell, then

$$\int_{\text{all k-space}} v^*(\mathbf{k})v(\mathbf{k})\,d\mathbf{k} = \Sigma_{\mathbf{K}_n}\int_{\mathrm{BZ}} |v(\mathbf{k}-\mathbf{K}_n)|^2\,d\mathbf{k}$$
$$= \frac{1}{\Omega}\int_{\mathrm{BZ}} d\mathbf{k} = \frac{\Omega_{\mathrm{B}}}{\Omega} = \frac{8\pi^3}{\Omega^2} \qquad (3.14)$$

where $\Omega_B$ is the volume of Brillouin zone. In the reduced zone scheme, separate $v(\mathbf{k} - \mathbf{K}_n)$ are considered; they are equal in pairs at the zone boundary. In the one-dimensional case (Fig. 3.1), the intercepts of any vertical line $k = k_s$ with the various curves give the coefficients $v(k_s - \mathbf{K}_n)$ (Fig. 3.1).

Each band is represented by a different $v$-function, still provided the bands are separated. If the band is narrow and deviates little from an atomic level, $v$ approaches the probability amplitude for momentum for the corresponding atomic state, decaying with increasing $\mathbf{k}$ as an inverse power. If we have an empty lattice, the wavefunctions are plane waves: only one of the coefficients is different from zero and is constant throughout a Brillouin zone. Wavefunctions such as that of the 3$s$ electron of sodium have two types of oscillations: the short wavelength oscillations near the nuclei, and the long wave $\lambda = 2\pi/|k|$, extending through the crystal. The former give $v$-components of high momenta, the latter a single component taken to be in the first Brillouin zone. This single component has a larger amplitude than the others. Curves of equal $v(\mathbf{k})$ in the plane $k_z = 0$, for the conduction band of sodium, are shown in Fig. 3.2 (Gumbrell, private communication). The circles indicate the regions occupied by the Fermi distribution.

Slater (1965) has calculated the one-dimensional momentum eigenfunctions for a one-dimensional lattice with a cosine potential. Fig. 3.3 shows the momentum eigenfunctions $v_1$, $v_2$ for the two lowest bands for a weak potential: each $v$ is approximately confined to one zone and almost constant in it. Fig. 3.4 shows the momentum eigenfunctions for the lowest seven bands for a strong potential: the lowest bands are practically narrow levels in the potential wells. The behaviour of Fig. 3.3 would be reproduced only for much higher bands. It is interesting that even for almost free electrons these $v(k)$ are not at all smooth functions. The $v$'s contain an arbitrary phase factor which ordinarily is chosen so that $v$ is real, or pure imaginary. Successive bands are alternatively even and odd functions of $k$.

In three dimensions these functions are very complicated and we illustrate briefly only the case of the simple cubic empty lattice. The problem is separable: $v(\mathbf{k}) = v_i(k_x)v_j(k_y)v_l(k_z)$. The first four Brillouin zones are given in Fig. 2.15. Each $v_i$ function is like those of Fig. 3.3 except that it is rigorously confined to the appropriate zone. The lowest three-dimensional band has a momentum eigenfunction which is just the product $v_1(k_x)v_1(k_y)v_1(k_z)$, so that $v_1(\mathbf{k})$ is confined to the first Brillouin zone (the central cube), where it is constant. For the second energy band, to which we associate $v_2$, complications arise because of degeneracies. Taking any point $\mathbf{k}$ not on a symmetry line, the energy $E_2$ (Eq. (2.44)) is the lowest of the six values.

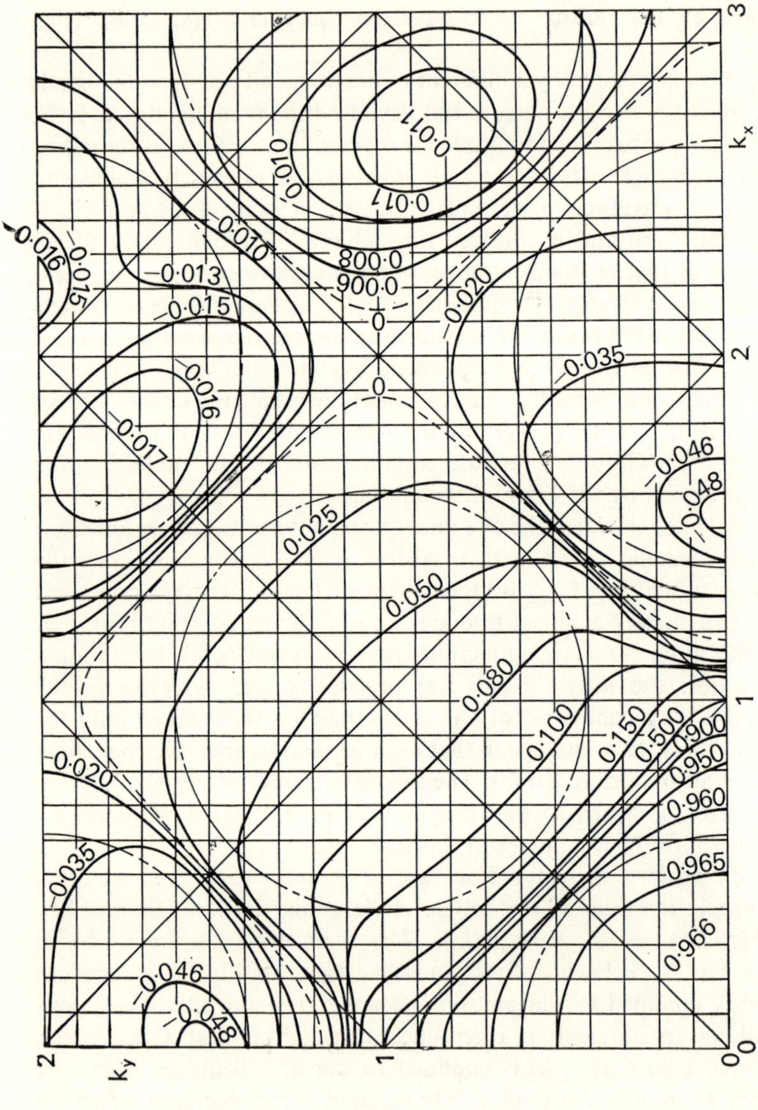


Fig. 3.2. Curves of constant momentum eigenfunction in a coordinate plane, for the conduction band of sodium. (After Gumbrell, private communication). The unit for $k$ is $2\pi/a$, where $a$ is the side of the elementary cube.

$$(k_x - 2)^2 + k_y^2 + k_z^2 \qquad (k_x + 2)^2 + k_y^2 + k_z^2$$
$$k_x^2 + (k_y - 2)^2 + k_z^2 \qquad k_x^2 + (k_y + 2)^2 + k_z^2$$
$$k_x^2 + k_y^2 + (k_z - 2)^2 \qquad k_x^2 + k_y^2 + (k_z + 2)^2$$

and which one it is, depends on the position of **k**. The other values correspond to higher bands. Here the side of the cube has been taken as $\pi$, so that the Brillouin zone extends, along each of the three axes, from $-1$ to $+1$. By

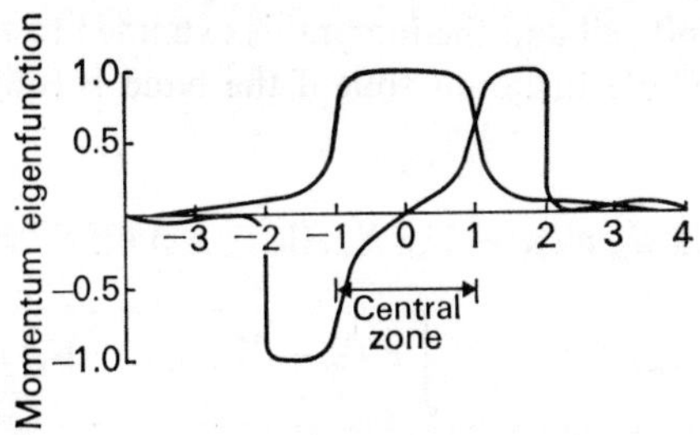


Fig. 3.3. Momentum eigenfunctions for the two lowest bands in a weak one-dimensional cosine potential. (After Slater, 1965)

inspection one recognises that $v_2$ is confined to the second Brillouin zone, that is to the region between the central cube and the circumscribed dodecahedron (Fig. 2.15). $v_2$ is in every sextant the product of two one-dimensional $v_1$-functions (along the faces of the central cube) and one one-dimensional

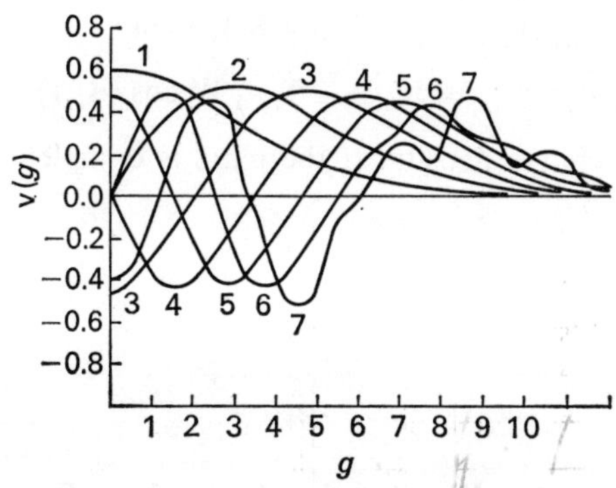


Fig. 3.4. Momentum eigenfunctions for the seven lowest bands in a strong one-dimensional cosine potential. (After Slater, 1965)

$v_2$-function (perpendicularly to these faces). Similarly $v_3$ is confined to the third Brillouin zone, and so on. In fact, consideration of the free electron bands (equation 2.44) is the simplest unambiguous method of constructing the various Brillouin zones through the rule that the band of index $n$ must occupy the $n$th-zone. For an example, see Sommerfeld and Bethe, 1933, page 392. If the electrons are not completely free, the $v$-functions are not strictly confined to one zone, but somewhat smeared out, as in Fig. 3.2.

The total charge density due to the electrons in one band, $\rho(\mathbf{r})$, is also a periodic function and can be expressed through the overlap integrals of the momentum eigenfunctions. We have $\rho(\mathbf{r}) = \int \Psi^*(\mathbf{r},\mathbf{k})\Psi(\mathbf{r},\mathbf{k})\, d\mathbf{k}$, where $\Psi$ is normalised in the unit cell and the integral is extended to the occupied region of $\mathbf{k}$-space (to the whole Brillouin zone if the band is fully occupied). Using (3.9) we obtain

$$\rho(\mathbf{r}) = \int d\mathbf{k}\, \Sigma_m\, v^*(\mathbf{k} - \mathbf{K}_m)\, \Sigma_n\, v(\mathbf{k} - \mathbf{K}_n)\, e^{i(\mathbf{K}_n - \mathbf{K}_m)\cdot\mathbf{r}}$$

$$= \Sigma_m\, \Sigma_n\, e^{i(\mathbf{K}_n - \mathbf{K}_m)\cdot\mathbf{r}} \int v^*(\mathbf{k} - \mathbf{K}_m) v(\mathbf{k} - \mathbf{K}_n)\, d\mathbf{k}$$

Denote by $W_{mn}$ the normalised overlap integrals of the momentum eigenfunctions

$$W_{mn} = \frac{1}{\Omega_{\mathrm{B}}} \int v^*(\mathbf{k} - \mathbf{K}_m) v(\mathbf{k} - \mathbf{K}_n)\, d\mathbf{k}$$

then

$$\rho(\mathbf{r}) = \Sigma_m\, \Sigma_n\, W_{mn}\, e^{i(\mathbf{K}_n - \mathbf{K}_m)\cdot\mathbf{r}}$$

We can collect all the terms in the sum for which $\mathbf{K}_n - \mathbf{K}_m = \mathbf{K}_i$ and introduce the symbol $W_i = \Sigma_{m,n}\, W_{mn}\ (\mathbf{K}_n - \mathbf{K}_m = \mathbf{K}_i)$; then

$$\rho(\mathbf{r}) = \Sigma_i\, W_i\, e^{i\mathbf{K}_i\cdot\mathbf{r}} = W_0 + 2\, \Sigma_{i\neq 0}\, W_i \cos(\mathbf{K}_i.\mathbf{r}) \tag{3.15}$$

$W_0$ is the average charge density. If the band is completely full, it reduces, because of (3.14), to

$$\frac{1}{\Omega_{\mathrm{B}}} \frac{\Omega_{\mathrm{B}}}{\Omega} = \frac{1}{\Omega}$$

(spin neglected).

## 3.3 Equations for the momentum eigenfunctions

The momentum eigenfunctions satisfy a system of algebraic equations, important because it is the basis of many methods of band structure calculations. Substituting (3.9) and (2.40) into Schrödinger's equation

$$-\nabla^2\Psi + V\Psi = E\Psi$$

we obtain

$$\Sigma_{\mathbf{K}_n} [|\mathbf{k} + \mathbf{K}_n|^2 - E] v(\mathbf{k} - \mathbf{K}_n)\, e^{i(\mathbf{k} + \mathbf{K}_n)\,.\,\mathbf{r}}$$
$$+ \Sigma_{\mathbf{K}_m} V(\mathbf{K}_m)\, \Sigma_{\mathbf{K}_n} v(\mathbf{k} - \mathbf{K}_n)\, e^{i(\mathbf{k} + \mathbf{K}_n + \mathbf{K}_m)\,.\,\mathbf{r}} = 0$$

Multiply by $\exp[-i(\mathbf{k} + \mathbf{K}_i).\mathbf{r}]$ and integrate over a unit cell; then, because of the orthogonality of the plane waves, the first sum reduces to the single term $\mathbf{K}_n = \mathbf{K}_i$ and in the second sum $\mathbf{K}_n + \mathbf{K}_m = \mathbf{K}_i$, so that

$$[|\mathbf{k} + \mathbf{K}_i|^2 - E] v(\mathbf{k} - \mathbf{K}_i)$$
$$+ \Sigma_{\mathbf{K}_m} V(\mathbf{K}_m) v(\mathbf{k} - \mathbf{K}_i + \mathbf{K}_m) = 0 \quad (3.16)$$

which is the required system of equations.

For instance, in one dimension, writing $n$ for $\mathbf{K}_n$

$$[(k + n)^2 - E] v(k - n) + \sum_{m=-\infty}^{+\infty} V_m v(k - n + m) = 0$$

$(k-n+m)\{[k+n)^2-E]\delta_{m0} + V_m\} = 0$

$$n = \ldots -1, 0, +1 \ldots \quad (3.17)$$

This system of homogeneous equations for the $v$s has non-zero solutions only if the determinant of the cofficients vanishes; for instance, in the case of (3.17)

$$\begin{vmatrix} \cdots & \cdot & \cdot & \cdot & \cdots \\ \cdots & (k-1)^2 + V_0 - E & V_{-1} & V_{-2} & \cdots \\ \cdots & V_1 & k^2 + V_0 - E & V_{-1} & \cdots \\ \cdots & V_2 & V_1 & (k+1)^2 + V_0 - E & \cdots \\ \cdots & \cdot & \cdot & \cdot & \cdots \end{vmatrix} = 0$$

(3.18)

This condition is a secular equation for $E$, determining the various $E_n$ for the chosen value of $\mathbf{k}$. Equation (3.18) is in fact a special case of (2.32), when the basis functions for the variational procedure are the plane waves $e^{i(\mathbf{k} + \mathbf{K}_n)\,.\,\mathbf{r}}$.

If $\mathbf{k}$ is a point of high symmetry, for example the origin of $\mathbf{k}$-space, many of the vectors $\mathbf{k} + \mathbf{K}_n$ are equivalent, and one must construct combinations of the equivalent plane waves having the appropriate symmetry (Section 6.2). Some of the $v$s become then equal, or simple multiples of one another, or disappear from the equations; the number of equations is thereby reduced. There is thus an advantage in carrying out calculations at these points. In any case (3.18) is an infinite determinant, so that the $E_n$ cannot be obtained exactly, but only to some approximation, either predetermined, or imposed by computational limitations.

The system of equations (3.16) can also be considered as Schrödinger's equation in $\mathbf{p}$-space. In fact, this has the general form

$$p^2\Psi(\mathbf{p}) + V\Psi(\mathbf{p}) = E\Psi(\mathbf{p}) \quad (3.19)$$

where $V$ must be expressed as an operator in **p**-space. Now, for our periodic potential we have the expansion (2.40) and, according to the rules of quantum mechanics, in each term we must replace $x$ by $i\partial/\partial p_x$, etc. We also have

$$e^{iK_{mx}i\partial/\partial p_x}\,\Psi(p_x) = \Psi(p_x - K_{mx})$$

so that (3.19) becomes

$$(p^2 - E)\Psi(\mathbf{p}) + \Sigma_{\mathbf{K}_m} V(\mathbf{K}_m)\Psi(\mathbf{p} - \mathbf{K}_m) \tag{3.20}$$

$|\Psi(\mathbf{p} - \mathbf{K}_m)|^2$, suitably normalised, is the probability of momentum $\mathbf{p} - \mathbf{K}_m$, thus (3.20), apart from a change of notation, is the same as (3.16).

As an example of application of (3.18), consider the numerical example of Section 2.3. The 'vectors' of the reciprocal lattice are $n\pi/3$. Taking the origin at the centre of the potential well, the Fourier expansion of the potential is

$$V = \frac{2}{9} - \frac{8}{9\pi}\left[\cos\frac{\pi}{3}x - \frac{1}{3}\cos\pi x + \frac{1}{5}\cos\frac{5\pi}{3}x - \ldots\right]$$

Thus at $k = 0$, considering only three $v$-coefficients, equation (3.18) is

$$\begin{vmatrix} \frac{\pi^2}{9} + \frac{2}{9} - E & -\frac{4}{9\pi} & 0 \\ -\frac{4}{9\pi} & \frac{2}{9} - E & -\frac{4}{9\pi} \\ 0 & -\frac{4}{9\pi} & \frac{\pi^2}{9} + \frac{2}{9} - E \end{vmatrix} = 0$$

with the roots $E_1 = 0{\cdot}1868$, $E_2 = 1{\cdot}319$, $E_3 = 1{\cdot}354$, which are a good approximation to the correct roots given in Table 2.3.1 With five terms the roots are $E_1 = 0{\cdot}1868$, $E_2 = 1{\cdot}308$, $E_3 = 1{\cdot}351$, $E_4 = 4{\cdot}61$, $E_5 = 4{\cdot}62$. At $k = \pi/6$, considering four coefficients, equation (3.18) is

$$\begin{vmatrix} \frac{\pi^2}{4} + \frac{2}{9} - E & -\frac{4}{9\pi} & 0 & +\frac{4}{27\pi} \\ -\frac{4}{9\pi} & \frac{\pi^2}{36} + \frac{2}{9} - E & -\frac{4}{9\pi} & 0 \\ 0 & -\frac{4}{9\pi} & \frac{\pi^2}{36} + \frac{2}{9} - E & -\frac{4}{9\pi} \\ +\frac{4}{27\pi} & 0 & -\frac{4}{9\pi} & \frac{\pi^2}{4} + \frac{2}{9} - E \end{vmatrix} = 0$$

with the roots $E_1 = 0{\cdot}3465$, $E_2 = 0{\cdot}6279$, $E_3 = 2{\cdot}652$, $E_4 = 2{\cdot}745$, which again are a good approximation.

Substituting any eigenvalue in the equations (3.17) we obtain the coefficients $v\left(k \pm \frac{n\pi}{3}\right)$ for the corresponding band. As we have seen, these can be represented by a single function $v_i(k)$ for each band $i$. Table 3.3.1 gives the $v$-function for the lowest band in our example. The function has been normalised so that $v^2\left(k \pm \frac{n\pi}{3}\right)$ is exactly the probability of momentum $\hbar\left(k \pm \frac{n\pi}{3}\right)$; for example an electron in state $k = \frac{\pi}{12}$ has probability $(0{\cdot}969)^2$ of having momentum $\hbar\frac{\pi}{12}$, $(0{\cdot}233)^2$ of momentum $\hbar\left(\frac{\pi}{12} - \frac{\pi}{3}\right)$, $(0{\cdot}081)^2$ of momentum $\hbar\left(\frac{\pi}{12} + \frac{\pi}{3}\right)$, etc.

**Table 3.3.1**

| $\pm k$ | $v(k)$ | $\pm k$ | $v(k)$ |
|---|---|---|---|
| 0 | 0·985 | $5\pi/12$ | 0·081 |
| $\pi/12$ | 0·969 | $\pi/2$ | 0·042 |
| $\pi/6$ | 0·706 | $7\pi/12$ | 0·009 |
| $\pi/4$ | 0·233 | $2\pi/3$ | 0·003 |
| $\pi/3$ | 0·123 | | |

If the origin had not been chosen at a point of inversion symmetry, the $v$-function would not have been real.

## 3.4 Wannier functions

We have seen that energy and Bloch functions have the periodicity of the reciprocal lattice, so we have the possibility of Fourier expansions in **k**-space. Here the basic periodic functions for the expansion are the 'plane waves in **k**-space' $e^{i\mathbf{R}_i \cdot \mathbf{k}}$ where $\mathbf{R}_i$ is any vector of the direct lattice. Thus, omitting the band index,

$$E(\mathbf{k}) = \Sigma_{\mathbf{R}_i} E(\mathbf{R}_i)\, e^{i\mathbf{R}_i \cdot \mathbf{k}} \tag{3.21}$$

with

$$E(\mathbf{R}_i) = \frac{1}{\Omega_B} \int_{\text{unit cell of k-space}} E(\mathbf{k})\, e^{-i\mathbf{R}_i \cdot \mathbf{k}}\, d\mathbf{k} \tag{3.22}$$

Since $E$ is an even function of $\mathbf{k}$, (3.21) is in fact a cosine series. The $E(\mathbf{R}_i)$ are called Fourier coefficients of the energy.

Developing now $\Psi$ as a periodic function of $\mathbf{k}$, consider first $\Psi$ at the origin; we have

$$\Psi(0, \mathbf{k}) = \Sigma_{\mathbf{R}_i} a(\mathbf{R}_i)\, e^{i\mathbf{R}_i.\mathbf{k}}$$

A similar development holds at any point $\mathbf{r}$. In complete analogy to the case of the momentum eigenfunctions, we write the coefficients as $a(\mathbf{r} - \mathbf{R}_i)$, thus obtaining

$$\Psi(\mathbf{r}, \mathbf{k}) = \Sigma_{\mathbf{R}_i} a(\mathbf{r} - \mathbf{R}_i)\, e^{i\mathbf{R}_i.\mathbf{k}} \tag{3.23}$$

Each coefficients is a function of $\mathbf{r}$ centred at the labelling $\mathbf{R}_i$-point. They are called Wannier functions (Wannier, 1937).

An expression of the type of (3.23) had been suggested by Bloch (1928) in his first papers on wavefunctions in crystals; he assumed that the $a(\mathbf{r} - \mathbf{R}_i)$ could be taken as atomic orbitals centred at $\mathbf{R}_i$ (see Section 5.5). Bloch's formula was an approximation; (3.23) is exact. It can be inverted in the usual way giving

$$a(\mathbf{r} - \mathbf{R}_i) = \frac{1}{\Omega_\mathrm{B}} \int \Psi(\mathbf{r}, \mathbf{k})\, e^{-i\mathbf{R}_i.\mathbf{k}}\, d\mathbf{k} \tag{3.24}$$

where the integration is extended to the unit cell of $\mathbf{k}$-space, which may be taken to be the Brillouin zone, of volume $\Omega_\mathrm{B}$, that is over all the states of one band irrespective of whether they are occupied or not. Equation (3.24) may be interpreted as a unitary transformation from the Bloch functions to the Wannier functions. The Bloch functions of only one band are used. That $a$ is a function only of the difference $\mathbf{r} - \mathbf{R}_i$ can easily be verified by adding any $\mathbf{R}_j$ to both arguments, that is going to an equivalent point in another cell; we have

$$\begin{aligned} a(\mathbf{r} + \mathbf{R}_j, \mathbf{R}_i + \mathbf{R}_j) &= \frac{1}{\Omega_\mathrm{B}} \int_\mathrm{BZ} \Psi(\mathbf{r} + \mathbf{R}_j, \mathbf{k})\, e^{-i\mathbf{R}_i.\mathbf{k}}\, e^{-i\mathbf{R}_j.\mathbf{k}}\, d\mathbf{k} \\ &= \frac{1}{\Omega_\mathrm{B}} \int_\mathrm{BZ} \Psi(\mathbf{r}, \mathbf{k})\, e^{-i\mathbf{R}_i.\mathbf{k}}\, d\mathbf{k} \\ &= a(\mathbf{r}, \mathbf{R}_i). \end{aligned}$$

The Wannier functions are localised since, for large $\mathbf{r} - \mathbf{R}_i$, the Bloch waves are bound to destroy each other by interference. It can be shown that they decay either exponentially, for non-degenerate bands, or as an inverse power. Thus, while the Bloch functions are appropriate when the position of the electron in the crystal is immaterial, the Wannier functions are useful when the position of the electron is of interest, for instance in the discussion of the motion of an electron near an impurity. In the case of bands deriving from deep atomic levels the Wannier function coincides practically with the atomic wavefunction. Any general wavefunction $\Psi$, for example a wave packet, can be expanded both in terms of Bloch functions and of Wannier functions.

In the same way as the momentum eigenfunctions, also the Wannier functions can be considered either as a set of coefficients $a(\mathbf{r} - \mathbf{R}_i)$ defined in one cell of the direct lattice, or as a single function $a(\mathbf{r})$ extending to the whole space. Fig. 3.5 gives a schematic representation of the Wannier functions

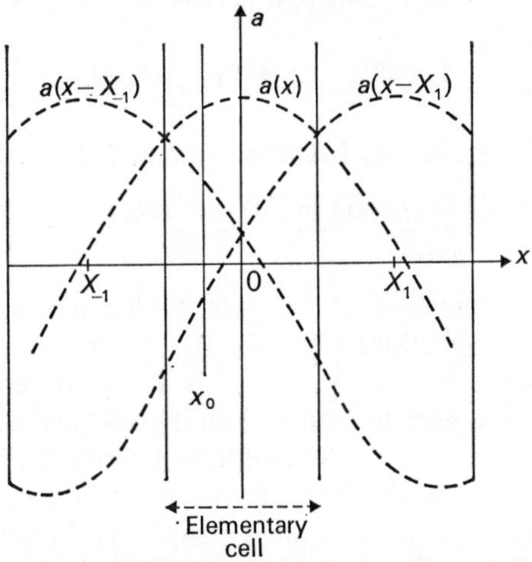


Fig. 3.5. Wannier functions in one dimension (schematic).

for a one-dimensional case: it is analogous to Fig. 3.1, with $x$ replacing $k$. The Wannier functions of one band are thus identical functions localised and centred at each lattice site. $a(\mathbf{r} - \mathbf{R}_i)$ is displaced by $\mathbf{R}_i$ with respect to $a(\mathbf{r})$. There is one Wannier function for each band, complications arising because of degeneracies, as with the momentum eigenfunctions. Wannier functions representing different bands are orthogonal.

As for the physical meaning of the $a$-coefficients, we have, for a completely full band,

$$\int_{\mathrm{BZ}} \Psi_n^*(\mathbf{k}, \mathbf{r})\Psi_n(\mathbf{k}, \mathbf{r})\, d\mathbf{k}$$

$$= \Sigma_{\mathbf{R}_i,\mathbf{R}_j}\, a_n^*(\mathbf{r} - \mathbf{R}_i)a_n(\mathbf{r} - \mathbf{R}_j)\int_{\mathrm{BZ}} e^{i(\mathbf{R}_j - \mathbf{R}_i).\mathbf{k}}\, d\mathbf{k}$$

$$= \Omega_{\mathrm{B}}\Sigma_{\mathbf{R}_i}\, a_n^*(\mathbf{r} - \mathbf{R}_i)\, a_n(\mathbf{r} - \mathbf{R}_i) \quad (3.25)$$

since the 'waves' $e^{i\mathbf{R}_j \cdot \mathbf{k}}$ are orthogonal within the Brillouin zone. Thus

$$\Omega_B \Sigma_{\mathbf{R}_i} |a_n(\mathbf{r} - \quad _i)|^2$$

gives the probability of finding an electron at $\mathbf{r}$, averaged over all the states of the $n$th band. (3.25) is equivalent to (3.15).

As an example, consider the lowest band in a simple cubic lattice of lattice spacing $a$, for the case of free electrons. Then

$$\Psi_{\mathbf{k}}(\mathbf{r}) = \frac{1}{\Omega^{\frac{1}{2}}} e^{i\mathbf{k} \cdot \mathbf{r}}$$

Since all the $a(\mathbf{r} - \mathbf{R}_i)$ are identical, it is only necessary to calculate $a(\mathbf{r})$

$$a(\mathbf{r}) = \frac{1}{\Omega_B} \frac{1}{\Omega^{\frac{1}{2}}} \int_{BZ} e^{i(k_x x + k_y y + k_z z)} \, dk_x \, dk_y \, dk_z \tag{3.26}$$

In a cubic lattice $k_x$, $k_y$, $k_z$ vary between $-\pi/a$ and $+\pi/a$. Thus

$$a(\mathbf{r}) = \frac{1}{\Omega^{\frac{1}{2}}} \frac{\sin(\pi x/a)}{\pi x/a} \frac{\sin(\pi y/a)}{\pi y/a} \frac{\sin(\pi z/a)}{\pi z/a} \tag{3.27}$$

$a(\mathbf{r})$ is normalised to one over whole space, a general property (see equation 3.30) of the Wannier functions when the Bloch functions are normalised in the unit cell. Since a free electron band is always degenerate at the band edge, $a(\mathbf{r})$ in this case goes to zero at infinity only according to an inverse power law. The oscillations of $a(\mathbf{r})$ ensure orthogonality, but have no direct physical meaning. In fact, the Wannier functions in general are convenient only for mathematical purposes. This is also shown by the fact that they are affected by the choice of the phases of the Bloch functions. This effect can be illustrated in the present example.

Consider for simplicity the one-dimensional case. If the phase of the Bloch functions, $\phi(k)$, is taken as zero in the whole Brillouin zone we have, as in (3.27),

$$a(x) = \text{const} \frac{\sin(\pi x/a)}{\pi x/a}$$

But we could, for instance, take $\phi = 0$ for negative $k$ and $\phi = \phi_0$ (constant) for positive $k$. Then $\left(k_0 = \frac{\pi}{a}\right)$

$$a(x) = \text{const} \frac{\sin \frac{1}{2} k_0 x}{k_0 x} \cos \tfrac{1}{2}(k_0 x + \phi_0) \, e^{\frac{1}{2} i \phi_0}$$

Each different law of variation of $\phi$ with $k$ produces a different Wannier function. It is easy to see that the most general admissable $\phi(\mathbf{k})$ consists of a periodic function of **k**, plus the scalar product of **k** with some lattice vector $\mathbf{R}_m$. Generally the phase is chosen so as to obtain the 'most concentrated' Wannier function. However, if one takes the constant $C$ in (2.18) to be 1, as one generally does, the Wannier function is real, and in most cases is also the most localised possible (see for instance Kohn 1959). Of course, the value of the sum in (3.25) is the same, whatever the choice of the phase.

Wannier functions are generally quite complicated. For instance, the first free-electron Wannier function for the face-centred cubic lattice is

$$a(x, y, z) = \frac{1}{(x^2 - y^2)(x^2 - z^2)}\left[\cos\frac{x}{2}(y \sin y + z \sin z) + x \sin\frac{x}{2}(\cos y + \cos z)\right]$$
$$+ \frac{1}{(y^2 - z^2)(y^2 - x^2)}\left[\cos\frac{y}{2}(z \sin z + x \sin x) + y \sin\frac{y}{2}(\cos z + \cos x)\right]$$
$$+ \frac{1}{(z^2 - x^2)(z^2 - y^2)}\left[\cos\frac{z}{2}(x \sin x + y \sin y) + z \sin\frac{z}{2}(\cos x + \cos y)\right]$$

For each band, there is a simple relation between the Wannier function $a_n(\mathbf{r})$ and the momentum eigenfunction $v_n(\mathbf{k})$. If in (3.24) we express $\Psi$ by means of (3.9), we obtain

$$a(\mathbf{r}) = \frac{1}{\Omega_B}\int_{BZ} d\mathbf{k}\, \Sigma_{\mathbf{K}_n}\, v(\mathbf{k} - \mathbf{K}_n)\, e^{i(\mathbf{k}+\mathbf{K}_n).\mathbf{r}}$$

The summation over all cells of reciprocal space coupled to the integration over one of these cells is equivalent to integration over the whole of **k**-space, thus it is immaterial whether the argument of $v$ is denoted by $\mathbf{k} - \mathbf{K}_n$ or by just **k**, and we have

$$a(\mathbf{r}) = \frac{1}{\Omega_B}\int_{\text{whole } \mathbf{k}\text{-space}} v(\mathbf{k})\, e^{i\mathbf{k}.\mathbf{r}}\, d\mathbf{k} \tag{3.28}$$

so that $a(\mathbf{r})$ is the Fourier integral transform of $v(\mathbf{k})$.

The converse of course is true, so that $a(\mathbf{r})$ and $v(\mathbf{k})$ form a pair of Fourier transforms.

$$v(\mathbf{k}) = \frac{1}{\Omega}\int_{\text{whole space}} a(\mathbf{r})\, e^{-i\mathbf{k}.\mathbf{r}}\, d\mathbf{r} \tag{3.29}$$

From (3.14), (3.28), (3.29) we have, because of the completeness relation for Fourier transforms,

$$\int_{\text{whole space}} |a(\mathbf{r})|^2 \, d\mathbf{r} = 1 \tag{3.30}$$

still provided the Bloch functions have been normalised in the unit cell.

The Wannier functions centred on different sites are orthogonal. In fact

$$\int_{\text{all space}} a^*(\mathbf{r} - \mathbf{R}_j)a(\mathbf{r} - \mathbf{R}_i) \, d\mathbf{r}$$

$$= \frac{1}{\Omega_B^2} \int_{\text{all space}} d\mathbf{r} \int_{\text{BZ}} \Psi^*(\mathbf{r}, \mathbf{k}) \, e^{i\mathbf{R}_j.\mathbf{k}} \, d\mathbf{k} \int_{\text{BZ}} \Psi(\mathbf{r}, \mathbf{k}') \, e^{-i\mathbf{R}_i.\mathbf{k}'} \, d\mathbf{k}'$$

$$= \frac{1}{\Omega_B^2} \int_{\text{BZ}} e^{i\mathbf{R}_j.\mathbf{k}} \, e^{-i\mathbf{R}_i.\mathbf{k}'} \, d\mathbf{k} \, d\mathbf{k}' \int_{\text{all space}} \Psi^*(\mathbf{r}, \mathbf{k})\Psi(\mathbf{r}, \mathbf{k}') \, d\mathbf{r}$$

Having imposed periodic boundary conditions, the last integral is in fact to be taken over the large parallelepiped containing $N$ cells used for the periodic boundary conditions; then the $\Psi$s corresponding to different values of $\mathbf{k}$ are orthogonal, so that, with the usual normalisation, the last integral is $N\delta(\mathbf{k} - \mathbf{k}')$. In the double integral over $\mathbf{k}$, $\mathbf{k}'$ we are now restricted to $\mathbf{k}' = \mathbf{k}$, so that the summation over $\mathbf{k}'$ has only one of the $N$ terms and we are left with

$$\int_{\text{all space}} a^*(\mathbf{r} - \mathbf{R}_j)a(\mathbf{r} - \mathbf{R}_i) \, d\mathbf{r}$$

$$= \frac{1}{\Omega_B} \int_{\text{BZ}} e^{i(\mathbf{R}_j - \mathbf{R}_i).\mathbf{k}} \, d\mathbf{k}$$

$$= \delta_{ij} \tag{3.31}$$

In the Bloch approximation, in which the Wannier functions in (3.23) are replaced by atomic orbitals, the atomic orbitals at different lattice sites are not however orthogonal.

Another property of the Wannier functions is that the matrix element of the one-particle Hamiltonian between two Wannier functions located at lattice sites $i$, $j$ is equal to the corresponding Fourier coefficient of the energy

$$H_{ij} = \int_{\text{all space}} a^*(\mathbf{r} - \mathbf{R}_i)(-\nabla^2 + V)a(\mathbf{r} - \mathbf{R}_j) \, d\mathbf{r}$$

$$= E(\mathbf{R}_j - \mathbf{R}_i) \tag{3.32}$$

where $E(\mathbf{R}_n)$ is defined by (3.22). $H$ has non-diagonal elements between Wannier functions because these are not solutions of Schrödinger's equation. To prove (3.22) consider

$$Ha(\mathbf{r} - \mathbf{R}_j) = \frac{1}{\Omega_\mathrm{B}} H \int_\mathrm{BZ} \Psi(\mathbf{r}, \mathbf{k})\, e^{-i\mathbf{R}_j\cdot\mathbf{k}}\, d\mathbf{k}$$

$H$ does not depend on $\mathbf{k}$, and

$$H\Psi(\mathbf{k}) = E(\mathbf{k})\Psi(\mathbf{k}),$$

thus

$$Ha(\mathbf{r} - \mathbf{R}_j) = \frac{1}{\Omega_\mathrm{B}} \int_\mathrm{BZ} \Psi(\mathbf{r}, \mathbf{k})\, e^{-i\mathbf{R}_j\cdot\mathbf{k}}\, E(\mathbf{k})\, d\mathbf{k} \qquad (3.33)$$

and

$$H_{ij} = \frac{1}{\Omega_\mathrm{B}^2} \int_\text{all space} d\mathbf{r} \int_\mathrm{BZ} \Psi^*(\mathbf{r}, \mathbf{k}')\, e^{i\mathbf{R}_i\cdot\mathbf{k}'}\, d\mathbf{k}'$$

$$\int_\mathrm{BZ} \Psi(\mathbf{r}, \mathbf{k})\, e^{-i\mathbf{R}_j\cdot\mathbf{k}}\, E(\mathbf{k})\, d\mathbf{k}$$

Again, inverting the order of integration, the integral with respect to $\mathbf{r}$ vanishes unless $\mathbf{k}' = \mathbf{k}$ and we are left with

$$H_{ij} = \frac{1}{\Omega_\mathrm{B}} \int_\mathrm{BZ} e^{i(\mathbf{R}_i - \mathbf{R}_j)\cdot\mathbf{k}}\, E(\mathbf{k})\, d\mathbf{k} = E(\mathbf{R}_j - \mathbf{R}_i) \qquad Q.E.D.$$

The matrix element depends only on the difference $\mathbf{R}_j - \mathbf{R}_i$. If the atoms are far apart, or the Wannier functions strongly localised, only the diagonal element $E(0)$ is appreciably different from zero. This element gives the average energy, and the fact that the other Fourier coefficients of $E$ are small means that the band is narrow. $E(0)$ is practically equal to the energy of the atomic level from which the band derives, and the Wannier function approaches the atomic orbital.

If in (3.33) we express $\Psi$ through (3.23), we obtain the system of differential equations satisfied by the Wannier functions

$$Ha(\mathbf{r} - \mathbf{R}_j) = \frac{1}{\Omega_\mathrm{B}} \int_\mathrm{BZ} \Sigma_{\mathbf{R}_i}\, a(\mathbf{r} - \mathbf{R}_i)\, e^{i(\mathbf{R}_i - \mathbf{R}_j)\cdot\mathbf{k}}\, E(\mathbf{k})\, d\mathbf{k}$$

and using (3.22)

$$\begin{aligned} Ha(\mathbf{r} - \mathbf{R}_j) &= \Sigma_{\mathbf{R}_i}\, a(\mathbf{r} - \quad {}_i)E(\mathbf{R}_i - \mathbf{R}_j) \\ &= \Sigma_{\mathbf{R}_s}\, a(\mathbf{r} - \quad {}_j - \quad {}_s)E(\mathbf{R}_s) \qquad (3.34) \end{aligned}$$

These equations may in principle be used instead of the Schrödinger equation, or instead of the equations (3.16), as a starting point for band structure calculations. It is however a complicated system, can be dealt with only by variational methods, and has not received much application. The $a$-function must be determined subject to (3.31). There are no matrix elements of $H$ between Wannier functions of different bands, except when degeneracies occur.

## 3.5 Representation theory in the band model

Any wave packet $\Psi(\mathbf{r})$ can be written as a superposition of Bloch functions, each denoted by a band index $n$ and a crystal momentum vector $\mathbf{k}$:

$$\Psi(\mathbf{r}) = \Sigma_n \int_{\mathrm{BZ}} f_n(\mathbf{k})\Psi_{n\mathbf{k}}(\mathbf{r})\, d\mathbf{k} \tag{3.35}$$

and $f_n(\mathbf{k})$ is then said to be the wavefunction in the crystal momentum, or Bloch, representation. The $f_n(\mathbf{k})$ can always be taken to have the periodicity of reciprocal space.

In this representation, which corresponds to the momentum representation of ordinary continuum quantum mechanics, the simplest operator is the crystal momentum itself, which we denote by $\mathbf{P}$, whose matrix is diagonal

$$\mathbf{P}_{nn'}(\mathbf{k}, \mathbf{k}') = \hbar\mathbf{k}\delta(\mathbf{k} - \mathbf{k}')\delta_{nn'} \tag{3.36}$$

This operator generates the lattice translations. To see this, we recollect that in continuum quantum mechanics the momentum operator $\mathbf{p}$ has the role of a generator of infinitesimal translations, that is, the operator

$$e^{i\varepsilon p_x} = 1 + i\varepsilon p_x = 1 + \hbar\varepsilon\frac{d}{dx} \qquad (\varepsilon \text{ infinitesimal})$$

applied to $\Psi(x)$ changes it into

$$\Psi(x) + \hbar\varepsilon\frac{d\Psi}{dx} = \Psi(x + \hbar\varepsilon)$$

In three dimensions, $1 + i\varepsilon\mathbf{n}.\mathbf{p}$ shifts any state by $\varepsilon\hbar$ in the direction of $\mathbf{n}$. For an isolated system the Hamiltonian commutes with all translations, and thus the momentum is a constant of the motion.

In a crystal, however, it is not $\mathbf{p}$ which is a constant of the motion, but the crystal momentum $\mathbf{P}$. If we apply the operator $e^{i\mathbf{P}.\mathbf{R}/\hbar}$, where for the moment $\mathbf{R}$ is any vector, to

$$\Psi_{n\mathbf{k}}(\mathbf{r}) = u_{n\mathbf{k}}(\mathbf{r})\, e^{i\mathbf{k}.\mathbf{r}}$$

we obtain $u_{n\mathbf{k}}(\mathbf{r})\, e^{i\mathbf{k}.(\mathbf{r}+\mathbf{R})}$, since, for any function of $\mathbf{P}$,

$$F(\mathbf{P})\Psi = F(\hbar\mathbf{k})\Psi$$

The exponential factor of the Bloch function has been shifted by **R**, while the periodic factor has remained the same. The new function bears no simple relation to the original one, except when **R** is a lattice translation $\mathbf{R}_i$, because then $u_{n\mathbf{k}}(\mathbf{r}) = u_{n\mathbf{k}}(\mathbf{r} + \mathbf{R}_i)$, so that

$$e^{i\mathbf{P}.\mathbf{R}_i/\hbar}\,\Psi(\mathbf{r}) = \Psi(\mathbf{r} + \mathbf{R}_i) \tag{3.37}$$

The crystal momentum operator generates the lattice translations.

The true momentum has matrix elements

$$\mathbf{p}_{nn'}(\mathbf{k}, \mathbf{k}') = \int \Psi^*_{n\mathbf{k}}(\mathbf{r})(-i\hbar\nabla)\Psi_{n'\mathbf{k}'}(\mathbf{r})\, d\mathbf{r}$$

the integration being over the large parallelepiped used for the periodic boundary conditions. Because of the orthogonality of the Bloch functions, this reduces to

$$\mathbf{p}_{nn'}(\mathbf{k}, \mathbf{k}') = \delta(\mathbf{k} - \mathbf{k}')(\hbar\mathbf{k}\delta_{nn'} - i\hbar \int u^*_n \nabla u_{n'}\, d\mathbf{r}) \tag{3.38}$$

The velocity **w** has matrix elements

$$\mathbf{w}_{nn'}(\mathbf{k}, \mathbf{k}') = \frac{1}{m_0} p_{nn'}(\mathbf{k}, \mathbf{k}') \tag{3.39}$$

and thus is also diagonal in **k**, but has interband matrix elements. The momentum operator (3.38) is associated with the **k**-dependent Hamiltonian given by (2.14).

The representation of any coordinate $x$ is more complicated and reference is made to an excellent review by Blount (1962) of the whole formalism of band theory. We quote only the result

$$\mathbf{r}f_n(\mathbf{k}) = i\,\mathrm{grad}_{\mathbf{k}}\, f_n + \Sigma_{n'}\, X_{nn'}(\mathbf{k}) f_{n'}(\mathbf{k})$$

where

$$\mathbf{X}_{n'n}(\mathbf{k}) = i \int u^*_{n'\mathbf{k}} \nabla_{\mathbf{k}} u_{n\mathbf{k}}\, d\mathbf{r}$$

These quantities are however sensitive to the choice of the phases of the Bloch functions. The operators **w**, **r** satisfy of course

$$\mathbf{w} = -\frac{i}{\hbar}[\mathbf{r}, H]$$

The representation which is the analogue of the coordinate representation of ordinary quantum mechanics is called the Wannier representation*. Its study can be approached by asking whether there exists an operator **R** canonically conjugate to **P** and, if so, what its properties are. First, we may expect that **R**

* A mixed representation, based on the translation properties in **k**-space, besides those in ordinary space, has been suggested by Zak (1970).

will have discrete eigenvalues because the eigenvalues of **P** are taken to be confined to the first Brillouin zone. In the same way, the fact that the angle of rotation about any axis is confined to the range 0 to $2\pi$ leads to the conjugate variable, the component of angular momentum along the axis, to have discrete eigenvalues. Since **P** commutes with the band index $n$, it is convenient that **R** should do the same, that is that its eigenfunctions can be constructed from Bloch functions of one band only. **R** must have properties similar to those of the position operator **r** of continuum mechanics. The eigenvalues of **r** are a continuum extending from $-\infty$ to $+\infty$, the eigenfunctions are $\delta$-functions. Since the eigenvalues of **R** are discrete, its eigenfunctions will not be $\delta$-functions, but will be localised. Now, the ordinary momentum operator generates infinitesimal translations, and the continuum so generated is the continuum of the eigenvalues of the operator **r** canonically conjugated to **p**. Similarly, the lattice translations $\mathbf{R}_i$, generated by the crystal momentum **P** are the eigenvalues of the operator **R** canonically conjugated to **P**.

As for the eigenfunctions of **R**, satisfying

$$\mathbf{R}a_{n,\mathbf{R}_i} = \mathbf{R}_i a_{n,\mathbf{R}_i} \tag{3.40}$$

they must have the following properties

1. They must be centred at the corresponding $\mathbf{R}_i$;
2. Functions centred at different $\mathbf{R}_i$ must be orthogonal;
3. They must be all the same functions of **r**, that is

$$a_{n,\mathbf{R}_i}(\mathbf{r} + \mathbf{R}_j) = a_{n,\mathbf{R}_i - \mathbf{R}_j}(\mathbf{r}) \tag{3.41}$$

These are all properties of the Wannier functions. To complete the identification, it can be proved (Blount, 1962) that, if the $a_{n,\mathbf{R}_i}$ defined by (3.40) are expanded in terms of the Bloch functions of the same band, the expansion is just (3.24).

The Wannier functions are thus the basis of the coordinate representation. This representation, from (3.23), (3.24), is obtained from the crystal momentum representation by making the unitary transformation $e^{i\mathbf{k}\cdot\mathbf{R}_j}$. A wave packet will be expressed as

$$\Psi(\mathbf{r}) = \Sigma_{i,n}\, g_n(\mathbf{R}_i) a_n(\mathbf{r} - \mathbf{R}_i) \tag{3.42}$$

where

$$g_n(\mathbf{R}_i) = \int_{\mathrm{BZ}} f_n(\mathbf{k})\, e^{i\mathbf{k}\cdot\mathbf{R}_i}\, d\mathbf{k} \tag{3.43}$$

Both the crystal momentum and the Wannier representations are useful in perturbation problems. Suppose we are concerned with solving the perturbed Schrödinger equation

$$(H_0 + H')\Psi = i\hbar\frac{\partial\Psi}{\partial t} \tag{3.44}$$

where $H_0$ is the unperturbed Hamiltonian containing the periodic potential, and $H'$ is a perturbation, for instance due to an impurity atom, or a dislocation line, or an external field. By using the Bloch eigenfunctions of the unperturbed Hamiltonian, $\Psi_n(\mathbf{k}, \mathbf{r})$, satisfying

$$H_0\Psi_n = E_n\Psi_n \tag{3.45}$$

we can expand $\Psi$ as

$$\Psi(\mathbf{r}, t) = \Sigma_n \int_{\text{one zone}} f_n(\mathbf{k}, t)\Psi_n(\mathbf{k}, \mathbf{r})\, d\mathbf{k} \tag{3.46}$$

The expansion coefficients $f_n(\mathbf{k}, t)$ satisfy the integral equation, found by substituting (3.46) into (3.44), multiplying by $\Psi^*_{n'}(\mathbf{k}', \mathbf{r})$, integrating over the whole crystal, and using (3.45)

$$\left[E_n(\mathbf{k}) - i\hbar\frac{\partial}{\partial t}\right]f_n(\mathbf{k}) + \Sigma_{n'}\int_{\mathrm{BZ}} \langle n\mathbf{k}|H'|n'\mathbf{k}'\rangle f_{n'}(\mathbf{k}')\, d\mathbf{k}' = 0 \tag{3.47}$$

where

$$\langle n\mathbf{k}|H'|n'\mathbf{k}'\rangle = \int_{\text{all space}} \Psi^*_n(\mathbf{k}, \mathbf{r})H'(\mathbf{r})\Psi_{n'}(\mathbf{k}', \mathbf{r})\, d\mathbf{r} \tag{3.48}$$

If, instead of using the Bloch functions, we use the Wannier functions, then we expand $\Psi$ as

$$\Psi(\mathbf{r}, t) = \Sigma_{i,n}\, g_n(\mathbf{R}_i, t)a_n(\mathbf{r} - \mathbf{R}_i) \tag{3.49}$$

This expansion, in terms of localised functions, is more useful for localised perturbations. The equation satisfied by the coefficients is obtained by substituting (3.49) into (3.44), multiplying by $a^*_m(\mathbf{r} - \mathbf{R}_j)$, integrating over the whole crystal and using the equations (3.34) satisfied by the Wannier functions and their conditions of orthogonality. The result is

$$\Sigma_{i,m}\left\{\left[E_m(\mathbf{R}_j - \mathbf{R}_i) - i\hbar\delta_{ij}\frac{\delta}{\partial t}\right]\delta_{nm} + H'_{n,m}(\mathbf{R}_j, \mathbf{R}_i)\right\}g_m(\mathbf{R}_i) = 0 \tag{3.50}$$

where

$$H'_{n,m}(\mathbf{R}_j, \mathbf{R}_i) = \int_{\text{all space}} a^*_n(\mathbf{r} - \mathbf{R}_j)H'(\mathbf{r})a_m(\mathbf{r} - \mathbf{R}_i)\, d\mathbf{r} \tag{3.51}$$

The equations (3.50), which are much on the lines of the valence bond approach, are difference equations, but Wannier made the fruitful suggestion to consider $\mathbf{R}$ as a continuous variable and the $g_n$ as continuous functions of it. Then (3.42) becomes an integral expression perfectly analogous to (3.35);

it is evident that the relation between **P**, **R** has become the same as that between **p**, **r** in continuum mechanics, so that we can write, in the representation in which **R** is diagonal

$$\mathbf{P} = -i\hbar \text{ grad} \qquad \mathbf{k} = -i \text{ grad} \tag{3.52}$$

and in all functions $F(\mathbf{P})$ we can replace **P** by $-i\hbar$ grad. However, care is required because after all we are differentiating with respect to a discrete variable. The question was discussed by Wannier (1959, p. 175) who showed that it is allowed to approximate the crystal momentum by a differential operator if

1. the operator $F(\mathbf{P})$ can be developed in a power series about a natural starting point, for example the edge of a band;
2. the function $f(\mathbf{r})$ on which $F(\mathbf{P})$ operates is, or can be made to be, a slowly varying function of **r**.

This approximation will be further discussed in connection with the equations of motion (Section 4.3). The procedure consists in using the **k**-dependent Hamiltonian (2.14) and in it replacing **k** by $-i$ grad. In this way, one sets in fact $\mathbf{P} = \mathbf{p}$ forgetting the indeterminacy of **k** in reciprocal space.

The matrix of **R** in the Bloch representation can also be constructed. It is found (see for instance Weinreich, 1965, p. 138) that the values of **k** which the operator **R** connects lie very close to each other, but that the diagonal elements of **R** are zero. The most important point is that, when the electron can be considered as well localised in the interior of the crystal, the operator **R** becomes $i \text{ grad}_{\mathbf{k}}$. Then by using

$$\dot{\mathbf{R}} = \frac{i}{\hbar}[H, \mathbf{R}]$$

one finds

$$\dot{\mathbf{R}}\Psi_{n\mathbf{k}}(\mathbf{r}) = \frac{i}{\hbar} \text{ grad}_{\mathbf{k}}\, E_n(\mathbf{k})\Psi_{n\mathbf{k}}(\mathbf{r}) \tag{3.53}$$

The eigenvalue of $\dot{\mathbf{R}}$ indicated by (3.53) is (Section 4.3) exactly the average velocity for a wave packet in the crystal.

Another type of expansion, useful for perturbation problems, was suggested by Luttinger and Kohn (1955) in terms of functions

$$\chi_n(\mathbf{k}, \mathbf{r}) = e^{i\mathbf{k}\cdot\mathbf{r}}\, u_n(\mathbf{k}_0, \mathbf{r}) \tag{3.54}$$

where $u_n(\mathbf{k}_0, \mathbf{r})$ is the periodic part of the Bloch function at a particular point $\mathbf{k}_0$, usually an extremum for band $n$. Also the $\chi_n$ form a complete orthonormal set.

# 4

# Dynamics of carriers in crystals

## 4.1 Effective potential

When an external potential $V(\mathbf{r})$, other than the periodic crystal potential, is applied to an electron in a crystal, the effective potential $U(\mathbf{r})$ acting on the electron, because of interactions with the other electrons, is not $V$, but $V$ divided by a quantity $\varepsilon(\mathbf{q}, \omega)$, conveniently considered as a dielectric constant, which is a function of the wave vector $\mathbf{q}$ and of the frequency $\omega$ of the applied perturbation.

To see this, consider for simplicity a free electron gas. The proper way of treating the problem of electron interaction is through many-body theory, but the result stated above can be justified in an elementary way. Consider one Fourier component of the effective potential acting on the electron at $\mathbf{r}$ at time $t$

$$U_{\mathbf{q}} = U_0\, e^{i\mathbf{q}.\mathbf{r}}\, e^{-i\omega t}\, e^{\alpha t} \tag{4.1}$$

The factor $e^{\alpha t}$ ensures that $U_{\mathbf{q}}$ vanishes at $t = -\infty$, but the time $\alpha^{-1}$ is supposed to be very long. $U_{\mathbf{q}}$ is assumed to be small with respect to the kinetic energy of the electron, so that perturbation theory can be applied. The perturbation is also taken to be a real quantity, so that the complex conjugate expression should be added to (4.1) making $U_{\mathbf{q}}$ a Hermitian operator. The electron is described by the unperturbed wavefunction $e^{i(\mathbf{k}.\mathbf{r}-E_0 t)}$, normalised per unit volume;

$$E_0 = k^2 \tag{4.2}$$

is the unperturbed energy (in the usual units). The perturbation (4.1) mixes to this state only the state with wave vector $\mathbf{k} + \mathbf{q}$, so that the perturbed wavefunction is

$$\Psi(\mathbf{r}, t) = e^{i(\mathbf{k}.\mathbf{r}-E_0 t)} + b(t)\, e^{i[(\mathbf{k}+\mathbf{q}).\mathbf{r}-E_+ t]} \tag{4.3}$$

where

$$E_+ = |\mathbf{k} + \mathbf{q}|^2 \tag{4.4}$$

and $b$ is a small quantity.

Substituting (4.3) in the time dependent Schrödinger equation,

$$i\frac{\partial \Psi}{\partial t} = H\Psi$$

where

$$H = -\nabla^2 + U_{\mathbf{q}}$$

we have

$$E_0\, e^{i(\mathbf{k}.\mathbf{r}-E_0 t)} + \left(i\frac{db}{dt} + E_+ b\right) e^{i[(\mathbf{k}+\mathbf{q}).\mathbf{r}-E_+ t]}$$

$$= k^2\, e^{i(\mathbf{k}.\mathbf{r}-E_0 t)} + |\mathbf{k}+\mathbf{q}|^2 b\, e^{i[(\mathbf{k}+\mathbf{q}).\mathbf{r}-E_+ t]} + U_{\mathbf{q}}\, e^{i(\mathbf{k}.\mathbf{r}-E_0 t)} \tag{4.5}$$

where the product of the two small quantities $b$, $U_0$ has been neglected. Using (4.2) and (4.4), (4.5) simplifies to

$$i\frac{db}{dt} = U_{\mathbf{q}}\, e^{-i\mathbf{q}.\mathbf{r}}\, e^{i(E_+ - E_0)t}$$

or, using (4.1)

$$i\frac{db}{dt} = U_0\, e^{i[E_+ - E_0 - \omega]t + \alpha t}$$

Integration gives

$$b = \frac{U_0}{i}\int_{-\infty}^{t} e^{i[E_+ - E_0 - \omega]t + \alpha t}\, dt$$

$$= \frac{U_0\, e^{i[E_+ - E_0 - \omega]t + \alpha t}}{E_0 - E_+ + \omega + i\alpha}$$

and thus

$$\Psi_{\mathbf{k}}(t) = e^{-iE_0 t}\left[e^{i\mathbf{k}\cdot\mathbf{r}} + \frac{U_0\, e^{-i\omega t + \alpha t}\, e^{i(\mathbf{k}+\mathbf{q}).\mathbf{r}}}{E_0 - E_+ + \omega + i\alpha}\right] \tag{4.6}$$

Associated with the change in $\Psi_{\mathbf{k}}$, there will be a change $\delta\rho_{\mathbf{k}}$ in the charge density due to the electron with unperturbed wave vector **k**. As a further simplification, the crystal is treated as uniformly positively charged, this positive charge being overall neutralised by the electrons. The change in charge density due to all the electrons is

$$\delta\rho(\mathbf{r}, t) = e\,\Sigma_{\mathbf{k}}\,(|\Psi_{\mathbf{k}}|^2 - 1)$$

$$= e\,\Sigma_{\mathbf{k}}\, a_{\mathbf{k}+\mathbf{q}}(t)\, e^{i\mathbf{q}.\mathbf{r}} + a^*_{\mathbf{k}+\mathbf{q}}(t)\, e^{-i\mathbf{q}.\mathbf{r}}$$

where

$$a_{\mathbf{k}+\mathbf{q}} = \frac{U_0\, e^{-i\omega t + \alpha t}}{E_0 - E_+ + \omega + i\alpha}$$

and $e$ is the electronic charge; the summation is over all occupied electron states, and terms of order $a^2$ have been neglected. When we remember that the complex conjugate expression must be added to (4.1), we see that $\delta\rho$ is made up of four terms which can be rearranged to give

$$\delta\rho = e\,\Sigma_{\mathbf{k}}\left[\frac{U_0}{E_0 - E_+ + \omega + i\alpha} + \frac{U_0}{E_0 - E_- - \omega - i\alpha}\right] e^{-i\omega t+\alpha t}\, e^{i\mathbf{q}.\mathbf{r}} + \text{complex conjugate} \quad (4.7)$$

where $E_- = |\mathbf{k} - \mathbf{q}|^2$,

so that each Fourier component of the charge density depends on, and is proportional to, only the corresponding Fourier component of the perturbation.

Since the summation is only over occupied states, the probability of occupation $f_0$ (generally the Fermi-Dirac function) must be introduced, so that

$$\delta\rho = eU_0\,\Sigma_{\mathbf{k}}\left[\frac{f_0(\mathbf{k})}{E_0 - E_+ + \omega + i\alpha} + \frac{f_0(\mathbf{k})}{E_0 - E_- - \omega - i\alpha}\right] \times e^{-i\omega t+\alpha t}\, e^{i\mathbf{q}.\mathbf{r}} + \text{complex conjugate}$$

If in the second sum the state $\mathbf{k}$ is re-labelled $\mathbf{k} + \mathbf{q}$, the two denominators become equal and of opposite sign, thus

$$\delta\rho = eU_0\,\Sigma_{\mathbf{k}}\frac{f_0(\mathbf{k}) - f_0(\mathbf{k}+\mathbf{q})}{E_0 - E_+ + \omega + i\alpha}\, e^{i(\mathbf{q}.\mathbf{r}-\omega t)}\, e^{\alpha t} + \text{complex conjugate} \quad (4.8)$$

and the sum is now over all states.

The charge re-distribution created by the potential $U$ produces a potential energy field $\phi$ acting on the electrons given by

$$\nabla^2\phi = -4\pi e\delta\rho$$

It may be assumed that $\phi$ has the same space and time variation as $\delta\rho$:

$$\phi = \phi_0\, e^{i(\mathbf{q}.\mathbf{r}-\omega t)+\alpha t} + \text{complex conjugate}$$

Then

$$q^2\phi_0 = 4\pi e^2 U_0\,\Sigma_{\mathbf{k}}\frac{f_0(\mathbf{k}) - f_0(\mathbf{k}+\mathbf{q})}{E_0 - E_+ + \omega + i\alpha}$$

But this new potential $\phi$ is itself a perturbation on the original electron distribution. To make the calculation self-consistent, the assumed perturbation $U$, that is the total potential acting on the electron, should already have contained $\phi$. In other words,

$$U = V + \phi$$

where $V$ is the actual external potential. If

$$V = V_0\, e^{i(\mathbf{q}\cdot\mathbf{r}-\omega t)+\alpha t} + \text{complex conjugate}$$

then

$$U_0 = V_0 + U_0 \frac{4\pi e^2}{q^2} \Sigma_{\mathbf{k}} \frac{f_0(\mathbf{k}) - f_0(\mathbf{k}+\mathbf{q})}{E_0 - E_+ + \omega + i\alpha}$$

or

$$U_0 = \frac{V_0}{\varepsilon(\mathbf{q}, \omega)} \tag{4.9}$$

where

$$\varepsilon(\mathbf{q}, \omega) = 1 + \frac{4\pi e^2}{q^2} \Sigma_{\mathbf{k}} \frac{f_0(\mathbf{k}) - f_0(\mathbf{k}+\mathbf{q})}{E_+ - E_0 - \omega - i\alpha} \tag{4.10}$$

This proves the initial statement: the potential acting on the electron is not the applied potential $V$, but $V$ divided by a quantity $\varepsilon(\mathbf{q}, \omega)$ which behaves like a dielectric constant and which is a function of the wavelength and frequency of the applied perturbation. Note however that this division of $V$ by $\varepsilon$, to obtain $U$, applies to the Fourier components. The result has indeed been deduced for a single Fourier component. However, we are using here a linear approximation and we have in fact neglected terms such as $bU_0$, or $b^2$; thus we can sum the effects of different Fourier components. This approximation of treating the Fourier components independently is known as the random phase approximation. Thus, if the applied potential is

$$V(\mathbf{r}, t) = \iint V(\mathbf{q}, \omega)\, e^{i\mathbf{q}\cdot\mathbf{r}}\, e^{-i\omega t}\, d\mathbf{q}\, d\omega, \tag{4.11}$$

then

$$U(\mathbf{r}, t) = \iint \frac{V(\mathbf{q}, \omega)}{\varepsilon(\mathbf{q}, \omega)}\, e^{i\mathbf{q}\cdot\mathbf{r}}\, e^{-i\omega t}\, d\mathbf{q}\, d\omega \tag{4.12}$$

a formula due to Lindhard (1954).

Of particular interest is the static case. The static dielectric constant is obtained by putting $\omega = \alpha = 0$ in (4.10) and, going to the limit $q \to 0$, setting

$$E_+ - E_0 = |\mathbf{k}+\mathbf{q}|^2 - \mathbf{k}^2 \simeq \mathbf{q}.\text{grad}_{\mathbf{k}}\, E$$

and

$$f_0(\mathbf{k}) - f_0(\mathbf{k}+\mathbf{q}) \simeq -\mathbf{q}.\text{grad}_{\mathbf{k}}\, E \frac{\partial f_0}{\partial E}$$

The summation is changed into an integral, obtaining

$$\varepsilon(\mathbf{q}, 0) = 1 + \frac{4\pi e^2}{q^2}\int\left(-\frac{\partial f_0}{\partial E}\right) d\mathbf{k}\,\frac{1}{4\pi^3}$$

The factor $(4\pi^3)^{-1}$ takes into account the fact that there is this number of states per unit volume of reciprocal space.

We choose $d\mathbf{k}$ as for evaluating (2.51) and obtain, since $\dfrac{df_0}{dE}$ is a function of $E$ only,

$$\varepsilon(\mathbf{q}, 0) = 1 - \frac{4\pi e^2}{q^2}\int \frac{df_0}{dE}\frac{dn_s}{dE}\,dE$$

where $\dfrac{dn_s}{dE}$ is the density of states. But $df_0/dE$ is practically a negative $\delta$-function at the Fermi level, $E_F$, thus

$$\varepsilon(\mathbf{q}, 0) = 1 + \frac{4\pi e^2}{q^2}\frac{dn_s}{dE}(E_F) \tag{4.13}$$

We shall see (Section 5.2) that the density of states at the Fermi level, for electrons of density $n$, is $3n/2E_F$. (4.13) shows that, as $q \to 0$, $\varepsilon \to \infty$, $U \to 0$: an external field of long wavelength is screened out almost completely by the free electrons. Writing

$$2\sqrt{\pi}\, e\left[\frac{dn_s}{dE}(E_F)\right]^{\frac{1}{2}} = \lambda, \qquad \varepsilon(\mathbf{q}, 0) = 1 + \frac{\lambda^2}{\mathbf{q}^2} \tag{4.14}$$

the quantity $1/\lambda$ is called the screening distance, or screening radius.

## 4.2 Screening

A typical case of a perturbing field in a solid is the electrostatic field around a charged impurity. To deal with this, we need the Fourier transform of $(e^{-\lambda r})/r$. This is

$$\begin{aligned} F(q, \lambda) &= \int^{\infty} \frac{e^{-\lambda r}}{r} e^{i\mathbf{q}\cdot\mathbf{r}}\, d\mathbf{r} \\ &= 2\pi \int_0^{\infty} r\, dr\, e^{-\lambda r} \int_0^{\pi} e^{iqr\cos\theta} \sin\theta\, d\theta \\ &= 2\pi \int_0^{\infty} r\, e^{-\lambda r}\, dr \left[\frac{-e^{iqr\cos\theta}}{iqr}\right]_0^{\pi} \\ &= 4\pi \int_0^{\infty} r\, e^{-\lambda r}\, dr \frac{\sin qr}{qr} \end{aligned}$$

$$= \frac{4\pi}{q}\int_0^\infty e^{-\lambda r}\sin qr\,dr$$

$$= \frac{4\pi}{q}\,\mathrm{Im}\int_0^\infty e^{iqr-\lambda r}\,dr$$

$$= \frac{4\pi}{q}\,\mathrm{Im}\,\frac{1}{\lambda - iq}$$

$$= \frac{4\pi}{\lambda^2 + q^2}$$

In particular, the Fourier transform of $1/r$ is $4\pi/q^2$. Returning to our problem, the 'external' field is a 'bare' Coulomb potential $V = e^2/r$. Its Fourier transform is $V(q) = 4\pi e^2/q^2$, which is infinite at $q = 0$, corresponding to the infinite range of the Coulomb potential. Using then (4.9), (4.13), (4.14), we have

$$U(q) = \frac{V(q)}{\varepsilon(q)} = \frac{4\pi e^2}{q^2 + \lambda^2} \tag{4.15}$$

and this is the Fourier transform of the potential

$$\frac{e^2}{r}\,e^{-\lambda r}$$

Thus the potential around a charged impurity is a screened Coulomb potential. In the case of a non-degenerate semiconductor, where the screening distance is very large, the effective potential is obtained simply by dividing the external potential by the static dielectric constant.

Even in more complicated theories the strong long-range interaction between the electrons can be reduced to a screened Coulomb potential. Each electron behaves as if it had around it a positive charge cloud, corresponding to the exclusion of other electrons which would normally have been in the neighbourhood. However it still carries charge *e*. As mentioned in Section 1.1, this system is best described as an excitation of a certain field. But for the purpose of band theory we can treat it as a particle, with a particular relation between energy and momentum. These quasi-particles are stable near the Fermi level, but tend to break down away from the Fermi surface.

## 4.3 Equations of motion

First we derive an expression for the average value of the velocity in the crystal momentum representation, that is when the electron is described by a Bloch function

$$\Psi_{n\mathbf{k}}(\mathbf{r}) = u_{n\mathbf{k}}(\mathbf{r})\,e^{i\mathbf{k}\cdot\mathbf{r}}$$

We have

$$\langle \mathbf{w} \rangle_{n\mathbf{k}} = \frac{1}{m_0} \int u^*_{n\mathbf{k}}(\mathbf{r})\, e^{-i\mathbf{k}.\mathbf{r}}\, (-i\hbar\nabla) u_{n\mathbf{k}}(\mathbf{r})\, e^{i\mathbf{k}.\mathbf{r}} \tag{4.16}$$

where the integration is over the volume in which $\Psi$ is normalised. Carrying out the differentiation, and using (2.15),

$$\begin{aligned} \langle \mathbf{w} \rangle_{n\mathbf{k}} &= \frac{1}{m_0} \int u^*_{n\mathbf{k}}(\mathbf{r})(\hbar\mathbf{k} - i\hbar\nabla) u_{n\mathbf{k}}(\mathbf{r})\, d\mathbf{r} \\ &= \frac{1}{\hbar} \int u^*_{n\mathbf{k}}(\mathbf{r}) \operatorname{grad}_{\mathbf{k}} H(\mathbf{k}) u_{n\mathbf{k}}(\mathbf{r})\, d\mathbf{r} \end{aligned} \tag{4.17}$$

where $\operatorname{grad}_{\mathbf{k}}$ operates on $H$ only. But the average value of the gradient of an operator is the gradient of the eigenvalue, thus

$$\langle \mathbf{w} \rangle_{n\mathbf{k}} = \frac{1}{\hbar} \operatorname{grad}_{\mathbf{k}} E(\mathbf{k}) \tag{4.18}$$

For instance $w_x = \dfrac{1}{\hbar}\left(\dfrac{\partial E}{\partial k_x}\right)$, etc. If a face of the Brillouin zone is perpendicular to the $k_x$-axis, and the constant energy surfaces cut perpendicularly this face. then on this face $w_x = 0$. This expresses the fact that the Bragg reflection produces standing waves. We see that the velocity of an electron in state $\mathbf{k}$ is directed along the normal to the equi-energy surface through $\mathbf{k}$. Only for spherical surfaces this direction coincides everywhere with that of $\mathbf{k}$ itself.

The velocity (1.48) determines the electric current density $\mathbf{j} = -e\mathbf{w}$ (one electron per unit volume).

We know already that the velocity operator is diagonal in $\mathbf{k}$, but has interband matrix elements (cf. (3.38), (3.39)). These can generally be neglected when the bands are well separated.

If we superpose many Bloch waves corresponding to neighbouring $\mathbf{k}$-values and to the same band, we build a wave-packet. Since $\mathbf{w}$ is diagonal in $\mathbf{k}$, the average velocity of this wave-packet is the weighted mean of the velocities of the component waves. In practically all cases these have nearly the same velocity, so that the wave-packet also moves with the average velocity (4.18). However it spreads more and more with time.

Note that the proof of (4.18) holds only in the absence of a magnetic field, since a vector potential term was not included in the velocity operator.

Coming now to the equations of motion, it is convenient to discuss them in the coordinate representation, that is, to start from the equations (3.50) and use Wannier's approximation of treating the coefficients $g_n(\mathbf{R}_i, t)$ as continuous functions of $\mathbf{r}$. Then $g_n$ becomes a time-dependent envelope function for the

Wannier function $a_n$, and the system of difference equations (3.50) is replaced by differential equations.

To do this, the Fourier coefficients of the energy, $E_n(\mathbf{R}_s)$ must be changed into operators and we must establish what is their effect on an arbitrary function. Consider the expansion (3.21)

$$E_n(\mathbf{k}) = \Sigma_{\mathbf{R}_s} E_n(\ _s)\, e^{i\mathbf{R}_s\cdot\mathbf{k}}$$

and use the approximation (3.52) $\mathbf{k} = -i\nabla$. Then

$$E_n(-i\nabla) = \Sigma_{\mathbf{R}_s} E_n(\mathbf{R}_s)\, e^{\mathbf{R}_s\cdot\nabla} \tag{4.19}$$

But

$$\exp\left(R_x\frac{\partial}{\partial x}\right)f(x) = \left(1 + R_x\frac{\partial}{\partial x} + \frac{R_x^2}{2!}\frac{\partial^2}{\partial x^2} + \dots\right)f(x)$$

$$= f(x + R_x)$$

so that

$$E_n(-i\nabla)g_n(\mathbf{r}) = \Sigma_{\mathbf{R}_s}\, g_n(\mathbf{r} + \mathbf{R}_s)E_n(\mathbf{R}_s) \tag{4.20}$$

Now re-write (3.50) replacing $\mathbf{R}_i$ by $\mathbf{R}_s + \mathbf{R}_j$. The first term, since $E_n$ is an even function of the argument, becomes

$$\Sigma_s\, E_n(\mathbf{R}_s)g_n(\mathbf{R}_j + \mathbf{R}_s) = E_n(-i\nabla)g_n(\mathbf{r})_{r=\mathbf{R}_j}$$

Replace now $\mathbf{R}_j$ by the continuous variable $\mathbf{r}$; the result is

$$E_n(-i\nabla)g_n(\mathbf{r}, t) - i\hbar\frac{\partial g_n(\mathbf{r}, t)}{\partial t} + \Sigma_{i,m}\, H'_{nm}(\mathbf{R}_i, \mathbf{R}_j)g_m(\mathbf{R}_i, t) = 0 \tag{4.21}$$

which is the required equation of motion, determining the time variation of $\Psi$ in (3.49). In practice, (4.21) is used in conjunction with several approximations. First, wavefunctions out of only one band are used, that is, all matrix elements $H'_{nm}(\mathbf{R}_i, \mathbf{R}_j)$ are ignored when $n \neq m$. In physical terms, it is assumed that the perturbation is not strong enough to induce interband transitions. More important, one takes that $H'_{nm}(\mathbf{R}_i, \mathbf{R}_j)$ is zero unless $i = j$. This is equivalent to assuming that the perturbation varies slowly with position (as it must do for the Wannier approximation to be valid), so that $H'(\mathbf{R}_i, \mathbf{R}_j)$ vanishes because of the orthogonality of the wavefunctions. This approximation may not hold near a charged point impurity in a lattice.

Using these approximations, $H'_{nn}(\mathbf{R}_j, \mathbf{R}_j)$ can be written simply as $U(\mathbf{r})_{\mathbf{r}=\mathbf{R}_j}$ and then treated as an ordinary potential function of the continuous variable $\mathbf{r}$. The equation of motion becomes

$$\left[E_n(-i\nabla) - i\hbar\frac{\partial}{\partial t}\right]g_n(\mathbf{r}, t) + U(\mathbf{r})g_n(\mathbf{r}, t) = 0 \tag{4.22}$$

We see that $g_n$ plays the role of the wavefunction in an equation equivalent to the Schrödinger time equation.

The essential difference bettween (3.50) and (4.22) is that the unperturbed Hamiltonian $H_0$ in the starting equation (3.44) is not included explicitly. The effect of the field of the perfect crystal is included in the operator $E_n(-i\nabla)$, so that, if we know the dispersion relation $E(\mathbf{k})$ for a band, we can discuss all the dynamical effects of slow perturbing fields ignoring the lattice potential as such; we have instead a modified kinetic energy operator $E_n(-i\nabla)$.

Thus in any problem involving 'external' potentials we must solve the equivalent Schrödinger equation (4.22). If $E_n$ is a complicated function of $\mathbf{k}$, the mathematics may be difficult. Only for parabolic bands ($E = \hbar^2k^2/2m^*$) the operator is the usual Laplacian $\nabla^2$. If (conduction band of silicon)

$$E = A(k_x^2 + k_y^2) + Bk_z^2,$$

the operator has the form

$$a\left(\frac{\partial^2}{\partial x^2} + \frac{\partial^2}{\partial y^2}\right) + b\frac{\partial^2}{\partial z^2} \tag{4.23}$$

The Schrödinger equation (4.22) is associated to the classical Hamiltonian

$$H_{\text{cl}} = \Sigma_{\mathbf{R}_s} E_n(\mathbf{R}_s)\, e^{\mathbf{p}\cdot\mathbf{R}_s/\hbar} = E_n(\mathbf{p}) \tag{4.24}$$

The principle of correspondence can be applied: the wave packets follow the same trajectories as classical particles obeying Hamilton's equations of motion

$$\frac{d\mathbf{r}}{dt} = \text{grad}_{\mathbf{p}}\, H_{\text{cl}}; \qquad \frac{d\mathbf{p}}{dt} = -\text{grad}\, H_{\text{cl}}$$

The first of these equations give, with $\mathbf{p} = \hbar\mathbf{k}$,

$$\mathbf{w} = \frac{1}{\hbar}\,\text{grad}_{\mathbf{k}}\, E(\mathbf{k})$$

in agreement with (4.18). The second Hamilton equation gives

$$\hbar\frac{d\mathbf{k}}{dt} = -\text{grad}\, U(\mathbf{r}) \tag{4.25}$$

For instance, if $U$ is the potential energy in an electrostatic field $\mathscr{E}$, the right hand side is the force $e\mathscr{E}$. Thus crystal momentum obeys Newton's second law: it behaves like ordinary momentum as far as external forces are concerned.

If also a magnetic field $\mathbf{B}$ is present, it can be proved (see for instance

Slater 1967, p. 288) that the rate of change of crystal momentum is equal to the Lorentz force

$$\hbar\frac{d\mathbf{k}}{dt} = e\left(\mathscr{E} + \frac{1}{c}\mathbf{w} \times \mathbf{B}\right) \tag{4.26}$$

where **w** is given by (4.18). This is true provided transitions between different bands can be ignored.

## 4.4 Inverse mass tensor

It may be helpful to consider the dynamics of carriers also from the most elementary point of view. We have already observed that (4.18) is just the well known expression for the group velocity of a wave packet. In fact, when a wave $e^{i(kx-\omega t)}$ propagates in a dispersive medium, the phase velocity is $\omega/k$ and the group velocity is $d\omega/dk$, or, in three dimensions $\mathbf{v}_g = \text{grad}_\mathbf{k}\,\omega$. By the principle of correspondence, the velocity of the electron is identical to the group velocity of the associated waves. Since energy $E$ and momentum $\mathbf{p}$ are connected to $\omega$ and $\mathbf{k}$ through $E = \hbar\omega$, $\mathbf{p} = \hbar\mathbf{k}$, the particle velocity is

$$\mathbf{w} = \text{grad}_\mathbf{p}\, E = \frac{1}{\hbar}\,\text{grad}_\mathbf{k}\, E$$

If now an electron in a crystal is subjected to an applied force **F** (due to an electric field), the rate of change of its energy is

$$\frac{dE}{dt} = \mathbf{F}.\mathbf{w}$$

(the argument cannot be applied directly to a magnetic force which does no work); the change of energy in time $dt$ is then

$$dE = (\mathbf{F}.\mathbf{w})\,dt$$

but, considering $E$ as a function of **k**,

$$dE = \text{grad}_\mathbf{k}\, E.d\mathbf{k}$$

so that, using (4.18).

$$\frac{1}{\hbar}\mathbf{F}.\text{grad}_\mathbf{k}\, E\,dt = \text{grad}_\mathbf{k}\, E.d\mathbf{k}$$

or

$$\text{grad}_\mathbf{k}\, E.\left(\frac{1}{\hbar}\mathbf{F}\,dt - d\mathbf{k}\right) = 0$$

that is, in agreement with (4.25),

$$F = \hbar\frac{d\mathbf{k}}{dt} \tag{4.27}$$

It can be shown that (4.27) holds also at the critical points where $\text{grad}_\mathbf{k} E = 0$. In the absence of external forces, **k** is constant: an electron is not scattered by the atoms of a perfect lattice. This is in agreement with experiment, and the explanation of this fact was one of the early successes of quantum theory. The electron is scattered however by any imperfection of the lattice, and by the lattice vibrations. Typically, the mean free path at 300 K is $\sim 5 \times 10^{-6}$ cm in a metal like silver, and $\sim 10^{-4}$ cm in a semiconductor like indium antimonide. Clearly the electron 'goes through' a large number of atoms before being scattered.

Conservation of crystal momentum holds also for the collision of two electrons. A system of two electrons in a crystal is described by a wave-function $\Psi_\mathbf{k}(\mathbf{r}_1, \mathbf{r}_2)$ and when the lattice undergoes a translation $T_i$ by a lattice vector $\mathbf{R}_i$, we have

$$T_i\Psi_\mathbf{k}(\mathbf{r}_1, \mathbf{r}_2) = e^{i\mathbf{k}.\mathbf{R}_i}\,\Psi_\mathbf{k}(\mathbf{r}_1, \mathbf{r}_2) \tag{4.28}$$

where $\hbar\mathbf{k}$ is the crystal momentum for the system, a constant of the motion. But when the electrons are sufficiently far apart, both before and after the collision

$$\Psi_\mathbf{k}(\mathbf{r}_1, \mathbf{r}_2) = \Psi_{\mathbf{k}_1}(\mathbf{r}_1)\Psi_{\mathbf{k}_2}(\mathbf{r}_2)$$

and

$$T_i\Psi_\mathbf{k}(\mathbf{r}_1, \mathbf{r}_2) = e^{i(\mathbf{k}_1+\mathbf{k}_2).\mathbf{R}_i}\,\Psi_\mathbf{k}(\mathbf{r}_1, \mathbf{r}_2) \tag{4.29}$$

so that

$$\mathbf{k}_1 + \mathbf{k}_2 = \mathbf{k} = \text{constant} \tag{4.30}$$

However we know that (4.28), (4.29) are still valid if a reciprocal lattice vector **K** is added to any of the **k**s. Thus (4.30) must be written more generally

$$\mathbf{k}_1 + \mathbf{k}_2 + \mathbf{K}_n = \text{constant}$$

so that the sum of the crystal momenta of the two electrons after the collision is equal to the sum before the collision plus a reciprocal lattice vector. Collision processes in which $\mathbf{K}_n = 0$ are called normal, while processes involving a reciprocal lattice vector are called 'Umklappung' processes.

Coming now to the electron acceleration under an external force **F**, this is

$$\frac{d\mathbf{w}}{dt} = \frac{1}{\hbar}\,\text{grad}_\mathbf{k}\,\frac{dE}{dt}$$

$$= \frac{1}{\hbar}\,\text{grad}_\mathbf{k}\,(\mathbf{F}.\mathbf{w})$$

$$= \frac{1}{\hbar^2} \text{grad}_{\mathbf{k}} (\mathbf{F} . \text{grad}_{\mathbf{k}} E) \tag{4.31}$$

This equation is the expression, for electrons in crystals, of Newton's law of motion.

$$\frac{d\mathbf{w}}{dt} = \frac{1}{m}\mathbf{F}$$

It is valid provided the applied field is not too intense. In fact, in deriving it we have tacitly assumed that only one band need be considered. But in high fields ($\sim 10^5$ V cm$^{-1}$), electrons may be excited from one band to another (Zener effect) and (4.31) must then be generalised.

It is clear that the acceleration is not in general parallel to the applied force. It is not surprising that Newton's law should not apply in its simple form, since electrons in crystals are subjected to other forces, besides the external ones. The analogy with free electrons is made close by introducing the concept of effective mass $m^*$. The three components of the vector equation (4.31) are

$$a_i = \frac{d\mathbf{w}_i}{dt} = \frac{1}{\hbar^2} \sum_{j=1}^{3} \frac{\partial^2 E}{\partial k_i \partial k_j} F_j \tag{4.32}$$

so that, formally, the scalar $1/m$ has been replaced by a tensor (effective, or reciprocal mass tensor), of components

$$\left(\frac{1}{m^*}\right)_{ij} = \frac{1}{\hbar^2} \frac{\partial^2 E}{\partial k_i \partial k_j} \tag{4.33}$$

This reciprocal mass tensor, which has generally different values at different **k**-points, is a measure of the ease with which an electron is accelerated by an electric field, or, one may say, of the 'freedom' of the electron. The tensor $(m_0/m^*)$, where $m_0$ is the free electron mass is therefore sometimes called the freedom tensor, or $f$-tensor.

Thus we write

$$\mathbf{a} = \left(\frac{1}{m^*}\right)\mathbf{F} \tag{4.34}$$

In the special case of isotropic parabolic energy bands,

$$E = \text{constant}\,(k_x^2 + k_y^2 + k_z^2)$$

$m^*$ is a scalar. In one dimension

$$\frac{1}{m^*} = \frac{1}{\hbar^2} \frac{d^2 E}{dk^2} \Big/ m^* \tag{4.35}$$

$1/m^*$ is thus proportional to the curvature of the $E$ versus $k$ curve at any point. If the curvature is positive (the curve is convex towards the $k$-axis), the effective mass is positive, but, if the $E$ versus $k$ curve is concave towards the $k$-axis (negative curvature), the effective mass is negative. If the curvature is large, the effective mass is small, and vice versa.

In the three-dimensional case, if there is a band edge at $\mathbf{k} = 0$, the expansion of the energy holding for isotropic bands

$$E = E_0 + \frac{\hbar^2 k^2}{2m^*} + \ldots$$

is generalised to (*cfr* 2.46)

$$E = E_0 + \frac{\hbar^2}{2} \sum_{\alpha,\beta=x,y,z} \left(\frac{1}{m^*}\right)_{\alpha\beta} k_\alpha k_\beta + \ldots \tag{4.36}$$

an expression which is easily extended to cover the case of an edge at any arbitrary **k**-point.

When the inverse mass is a tensor, the effect of an electric field in accelerating an electron is different in different directions. If the constant energy surface on which the electron finds itself is elongated into a rotation ellipsoid (a 'valley'), the mass appears greater along the valley than in the transverse direction. This is important when calculating the average mobility of the electrons.

By a suitable choice of the coordinate axes, the $(1/m^*)$ tensor can always be diagonalized, that is reduced to the form

$$\begin{vmatrix} 1/m_x & 0 & 0 \\ 0 & 1/m_y & 0 \\ 0 & 0 & 1/m_z \end{vmatrix}$$

and thus the energy near an extremum can always be expressed by (2.48), that is

$$E(\mathbf{k}) = E(\mathbf{k}_0) + \tfrac{1}{2}\hbar^2\left[\frac{(k_x - k_{0x})^2}{m_x} + \frac{(k_y - k_{0y})^2}{m_y} + \frac{(k_z - k_{0z})^2}{m_z}\right] \tag{4.37}$$

The axes for which this happens are called the principal axes of the tensor, and the reciprocal values of the three diagonal elements are called the principal masses. Both the principal axes and the principal masses differ in general from one point to another of **k**-space. The principal masses are usually negative for states near the state, or states, of highest energy within each band.

A negative effective mass means that the acceleration is in opposite direction to the applied force. This startling state of affairs is a consequence of the use of the effective mass to express the effect of the lattice potential. In fact, the dynamical properties of the electron are influenced by the Bragg reflections and the consequent ability of the electron to transfer momentum to the lattice. A negative effective mass means that the momentum transfer is so large that an increase of **k** means a decrease of the velocity (Kittel, 1954). An example is discussed in connection with the nearly free electron approximation (Section 5.3). The properties of electrons with negative effective mass are so unlike those of ordinary particles, that it is much easier to discuss the properties of the absence of such an electron from a full band, that is of a positive hole (Section 4.8).

Another case in which the inverse mass tensor is not positive definite is when, as in copper, silver and gold, the Fermi surface makes contact with the Brillouin zone boundary (Section 5.2); then it can still be represented approximately by

$$E(\mathbf{k}) = \frac{\hbar^2 k_1^2}{2m_1} + \frac{\hbar^2 k_2^2}{2m_2} + \frac{\hbar^2 k_3^2}{2m_3}$$

but with $m_3$ negative, to give hyperbolic energy surfaces which can run through the zone boundary, in the repeated zone scheme. Some energy surfaces in Fig. 2.19c are of this type. Also the dynamics of these electrons is odd from a classical point of view, and they affect many properties, such as the magnetoresistance. What happens is that their motion is altered by the Bragg reflection from the zone boundary and they tend to move parallel to the planes of the lattice, as if by a series of wave guides.

An electric field that has the direction of one of the principal axes accelerates the electron as though it were free with a scalar mass equal to the corresponding principal mass, irrespective of the other two principal masses. The principal masses can thus also be called effective masses in the direction of the principal axes. They are the masses that a free electron would need in order that the velocity change under an applied impulse along that axis should equal the actual velocity change of the electron in the crystal. This concept of effective mass in a direction can be generalised. If **F** has an arbitrary direction, the acceleration is not in general parallel to it, but can be decomposed into two components, one parallel and one perpendicular to the field. Naturally, only the parallel component is of interest, as it is the quantity that determines the energy transfer from the field to the electron. The problem then arises of which scalar mass $m_F$ would have produced this parallel component of the acceleration. Now, if **e** is a unit vector in the direction of **F**, multiply (4.34) scalarly by **e** to obtain the required component $a_F = F/m_F$:

$$a_F = \mathbf{e}.\mathbf{a}$$

$$= \mathbf{e}.\left(\frac{1}{m^*}\right)\mathbf{F}$$

$$= F\mathbf{e}.\left(\frac{1}{m^*}\right)\mathbf{e}$$

$$= \frac{F}{m_F}$$

Therefore

$$\frac{1}{m_F} = \mathbf{e}.\left(\frac{1}{m^*}\right)\mathbf{e}$$

It is interesting to use the concept of effective mass to discuss what would be, in an absolutely perfect crystal, the motion of an electron through a band. Suppose that in the direction of an applied force $\mathbf{F}$ the $E - \mathbf{k}$ relation is as illustrated in Fig. 4.1. Consider an electron initially with $\mathbf{k} = 0$. Its acceleration is $e\mathbf{F}/m_F$; its velocity after time $dt$ is $e\mathbf{F}\,dt/m_F$, where $m_F$ is the effective mass

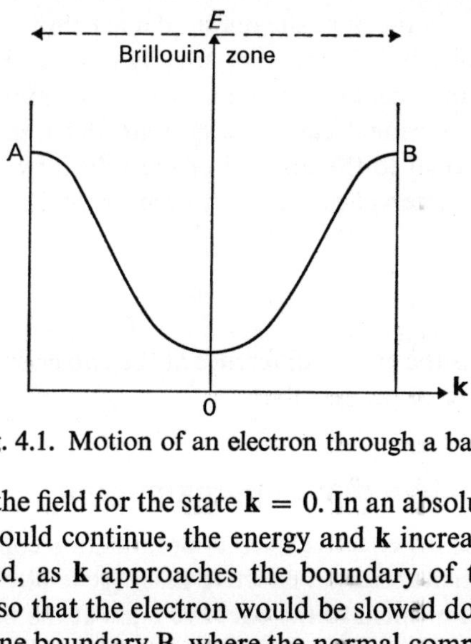


Fig. 4.1. Motion of an electron through a band.

in the direction of the field for the state $\mathbf{k} = 0$. In an absolutely perfect crystal this acceleration would continue, the energy and $\mathbf{k}$ increasing. However, $m_F$ changes with $\mathbf{k}$ and, as $\mathbf{k}$ approaches the boundary of the Brillouin zone, becomes negative, so that the electron would be slowed down, coming to rest when $\mathbf{k}$ is at the zone boundary B, where the normal component of $\mathbf{w}$ is zero. Since A and B are equivalent states, in the reduced zone scheme the electron would be considered to be at A and further acceleration would make it move from A towards 0, the direction of its velocity being reversed. This is just the Bragg reflection at the zone boundary. The electron would then oscillate and one can calculate that in ordinary space the amplitude of the oscillation would be typically of a fraction of a mm. The periodic motion is quantized but, except in extremely high fields, the separation of the energy levels is very small.

But, due to imperfections in the crystal and to the lattice vibrations, the actual mean free path is much smaller, so that the electron merely acquires a small drift velocity, producing an electric current, in the direction of the field, superposed to the much larger original velocity. Correspondingly, **k** does not change much, the position of the electron in the Brillouin zone shifts very little, and its energy is unchanged to first order. Thus it may be assumed that a band structure is unaltered in the presence of a not too intense electric field. Since

$$\hbar\frac{d\mathbf{k}}{dt} = e\mathscr{E}$$

if the average time between collisions is $\tau(\sim 10^{-12}\ \mathrm{s})$, the change in **k** is $e\mathscr{E}\tau/\hbar$, which, for a field of 1 V cm$^{-1}$, is $\sim 10^3\ \mathrm{cm}^{-1}$; the dimensions of the Brillouin zone are of the order of $10^8\ \mathrm{cm}^{-1}$.

If there is a partially filled band, all electrons are displaced in **k**-space in the direction of the field, so that (if $m^*$ is positive) there are more electrons moving in this direction than in the opposite one, and a net electric current arises. But if the band is completely full, we have already remarked that the current is always zero. This fundamental fact can now be made more evident. Take for simplicity a one-dimensional case; the current, determined by the average velocity, is proportional to the integral of the velocity with respect to $k$, integral taken over an interval of $k$ equal to the Brillouin zone. But

$$\mathbf{w} = \frac{1}{\hbar}\frac{d}{dk}$$

so that the integral is the energy difference at the two ends of the interval, and this vanishes because of periodicity.

## 4.5 The effective mass approximation

For semiconductors, the effective mass approximation consists in treating the carriers as free particles with an effective mass (generally the reciprocal of a tensor) equal for all of them to the effective mass at the band edge. The concept can also be used for metals, when the effective mass is an average over the Fermi surface. In this way one has a convenient and often suggestive way of describing experimental results. More precisely, suppose that a physical quantity $Q$ is given, for free electrons, by the expression

$$Q = f(m_0)$$

where $m_0$ is the free electron mass. If then for some real solid it is found experimentally to have the value $Q^*$, the corresponding effective mass is defined by

$$Q^* = f(m^*)$$

that is, $m^*$ is that value which, when introduced into the free electron formula, gives agreement with experiment. Clearly the usefulness of the concept depends on whether, for a given material, the same value of $m^*$ describes correctly several different empirical properties.

Now in most semiconductors this is true to a surprising degree, provided the properties in question do not involve rapidly varying fields. This implies that $E$ must be very nearly a quadratic function of the $k_i$ – equation (4.36) – higher terms being negligible. In these cases then, when studying the motion of carriers in an external field, the effect of the unperturbed periodic field may be expressed by an inverse mass tensor, whose elements are determined by the form of the $E(\mathbf{k})$ relation at the band edge. To find $m^*$ thus requires detailed knowledge of the dispersion relation near the edge, and conversely all information on the dispersion relation is expressed by $m^*$. In practice, the elements of the tensor are not calculated from a knowledge of the periodic potential, but deduced from measured properties, for example from cyclotron resonance experiments. This approximation is employed very widely because the characteristic properties of semiconductors involve relatively few electrons at the bottom of the conduction band and the absence of relatively few electrons (holes) at the top of the valence band.

Note that if

$$E = E_0 + \frac{\hbar^2 k^2}{2m^*}$$

then

$$\mathbf{w} = \frac{\hbar \mathbf{k}}{m^*}$$

and

$$\mathbf{p} = \frac{m_0}{m^*}\hbar \mathbf{k} \qquad (4.38)$$

a formula which provides the relation between true momentum and crystal momentum in the effective mass approximation.

An important example of a perturbing potential is that around a substitutional impurity in silicon or germanium,

$$U = \frac{e^2}{\varepsilon r} \qquad (\varepsilon = \text{dielectric constant})$$

An electron bound to such an impurity is described by the Schrödinger equation

$$-\frac{\hbar^2}{2m^*}\nabla^2\Psi - \frac{e^2}{\varepsilon r}\Psi = E\Psi \tag{4.39}$$

If $m^*$ could be treated as a scalar, (4.39) would be a Schrödinger equation for a hydrogen-like atom, leading to energy levels

$$E_n = -\frac{m^* e^4}{2\hbar^2\varepsilon^2 n^2} \tag{4.40}$$

If, for example, $\varepsilon = 10$, $m^* = 0{\cdot}1\ m_0$, the $E_n$s are $10^{-3}$ the corresponding values for the hydrogen atom. The Bohr radius $\varepsilon\hbar^2/m^*e^2$ is enlarged by a factor $\sim 100$. This justifies assuming that the potential is slowly varying in the region where the electron is found with greatest probability. Levels of the above order or magnitude are indeed observed, but the theory is more complicated, because $m^*$ is generally a tensor and then equations of the type of (4.23), or more complicated, must be used.

An expression for $m^*$ near an edge can be derived by pertubation theory. Assume for simplicity that the edge is at $k = 0$ and that the band is non-degenerate. Start from the differential equation (2.13) for the periodic part, $u_{n\mathbf{k}}(\mathbf{r})$ of a Bloch function. Writing

$$E_n'(\mathbf{k}) = E_n(\mathbf{k}) - \frac{\hbar^2 k^2}{2m_0}$$

$$H' = \frac{\hbar\mathbf{k}\cdot\mathbf{p}}{m_0} \tag{4.41}$$

equation (2.13) becomes

$$(H_0 + H')u_{n\mathbf{k}}(\mathbf{r}) = E_n'(\mathbf{k})u_{n\mathbf{k}}(\mathbf{r}) \tag{4.42}$$

where $H_0$ is the unperturbed Hamiltonian $-\dfrac{\hbar^2}{2m_0}\nabla^2 + V(\mathbf{r})$ and, for small values of $\mathbf{k}$, $H'$ can be treated as a perturbation ($\mathbf{k}\cdot\mathbf{p}$ approximation). Consider a set of unperturbed functions $u_{n0}$ at $\mathbf{k} = 0$, in which $n$ runs over all the bands; second order perturbation gives

$$E_n'(\mathbf{k}) = E_n'(0) + H'_{nn} + \Sigma_{n'\neq n}\frac{H'_{nn'}H'_{n'n}}{E_n'(0) - E_{n'}'(0)} \tag{4.43}$$

where

$$H'_{nn'} = \frac{\hbar\mathbf{k}\cdot\mathbf{p}_{nn'}(0)}{m_0} \tag{4.44}$$

$\mathbf{p}_{nn'}(\mathbf{k})$ being the matrix element of momentum

$$\mathbf{p}_{nn'}(\mathbf{k}) = -i\hbar \int_{\text{unit cell}} u_{n\mathbf{k}}(\mathbf{r}) \nabla u_{n'\mathbf{k}}(\mathbf{r})\, d\mathbf{r} \tag{4.45}$$

Observing that there are no linear terms in $\mathbf{k}$ because we assume that $\mathbf{k} = 0$ is an extremum, we have

$$E_n(\mathbf{k}) = E_n(0) + \frac{\hbar^2 k^2}{2m_0} + \frac{\hbar^2}{m_0^2} \Sigma_{n' \neq n} \frac{\mathbf{k}.\mathbf{p}_{nn'}(0)\mathbf{k}.\mathbf{p}_{n'n}(0)}{E_n(0) - E_{n'}(0)} \tag{4.46}$$

which has the required form (4.36) if we write

$$\left(\frac{1}{m_n^*}\right)_{\alpha\beta} = \frac{\delta_{\alpha\beta}}{m_0} + \frac{2}{m_0^2} \Sigma_{n' \neq n} \frac{p_{nn'}^{(\alpha)}(0) p_{n'n}^{(\beta)}(0)}{E_n(0) - E_{n'}(0)} \tag{4.47}$$

(a formula known as the sum rule for the effective mass, or '$f$' sum rule). For instance

$$\left(\frac{1}{m_n^*}\right)_{xy} = \frac{2}{m_0^2} \sum_{n' \neq n} \frac{\langle n|p_x|n'\rangle\langle n'|p_y|n\rangle}{E_n - E_{n'}}$$

the matrix elements and the energies being calculated at $\mathbf{k} = 0$.

The result is easily generalised to a non-degenerate band edge occurring at any arbitrary value $\mathbf{k}_0$ of $\mathbf{k}$. However, if the band edge is degenerate, it is not possible to write down an explicit expression for $E$ as a function of $\mathbf{k}$. If bands $n_1, n_2, \ldots, n_r$ all have edges at $\mathbf{k} = 0$, so that the edge is $r$-fold degenerate with energy $E_n(0)$, then the standard perturbation treatment for degenerate states (see for instance Schiff, 1949, p. 153) gives the energy in each band to order $k^2$ as one of the eigenvalues of the matrix.

$$\begin{vmatrix} E_n(0) + \dfrac{\hbar^2 k^2}{2m_0} + A_{n_1 n_1} & A_{n_1 n_2} & \cdots & A_{n_1 n_r} \\ A_{n_2 n_1} & E_n(0) + \dfrac{\hbar^2 k^2}{2m_0} + A_{n_2 n_2} & \cdots & A_{n_2 n_r} \\ \cdot & \cdot & \cdots & \cdot \\ A_{n_r n_1} & A_{n_r n_2} & \cdots & E_n(0) + \dfrac{h^2 k^2}{2m_0} + A_{n_r n_r} \end{vmatrix} \tag{4.48}$$

where

$$A_{n_i n_j} = \frac{\hbar^2}{m_0^2} \Sigma_{s \neq n} \frac{\mathbf{k}.\mathbf{p}_{n_i s}(0)\mathbf{k}.\mathbf{p}_{s n_j}(0)}{E_n(0) - E_s(0)} \tag{4.49}$$

This **k**.**p** method is the most useful theoretical procedure for predicting and analysing the details of the band structure of semi-conductors near band edges. The matrix elements of **p**, which would be difficult to calculate, are often obtained from suitable experimental data.

The form of the expression derived for the effective mass leads one to talk of its being produced by an interaction between bands, governed by the momentum matrix elements and the energy gap at the edge. The diagonal elements of the inverse mass tensor are

$$\left(\frac{m_0}{m_n^*}\right)_{\alpha\alpha} = 1 + \frac{2}{m_0}\Sigma_{n'} \frac{|p_{nn'}^{(\alpha)}|^2}{E_n(0) - E_{n'}(0)} \tag{4.50}$$

thus the interaction of a level $n$ with lower lying levels ($E_{n'} < E_n$) decreases the effective mass, while the interaction with higher states tends to increase it; however this last case generally applies to states at the top of a band, which have a negative effective mass. The absolute value (mass of the hole) is then decreased.

Generally for two bands edge close in energy and which occur at the same value of **k**, only the mutual interaction need be considered to account for their effective masses. In order to interact, the two bands must be of the same type (Section 6.1). Calling the bands 1 and 2 we have, after elimination of the interaction term from the two expressions (4.47)

$$\left(\frac{1}{m_1^*}\right)_{\alpha\beta} - \frac{\delta_{\alpha\beta}}{m_0} = -\left(\frac{1}{m_2^*}\right)_{\alpha\beta} + \frac{\delta_{\alpha\beta}}{m_0} \tag{4.51}$$

An example of two bands of this type is given by the light hole valence band of germanium and the conduction band minimum at $\mathbf{k} = 0$ (Section 6.5). These two bands are in many respects the mirror image of each other with respect to the midpoint of the energy gap. From experiment, the electron mass $m_e$ at $\mathbf{k} = 0$ is isotropic and $m_e = 0{\cdot}036\, m_0$, while the effective mass $m_v$ of an electron at the top of the light hole valence band is almost isotropic and $m_v = -0{\cdot}042\, m_0$. Note that $m_v$ is negative: the mass of the corresponding hole is positive. Then

$$\frac{1}{m_e} - \frac{1}{m_0} = \frac{27}{m_0}$$

and

$$-\frac{1}{m_v} + \frac{1}{m_0} = \frac{25}{m_0}$$

showing that the simple relation (4.51) is almost satisfied. It also follows from (4.51) that in cases of this type the hole mass must be larger than the electron mass.

As to the accuracy for the expression of $E_n(\mathbf{k})$ to terms or order $k^2$, successive terms have an extra factor $S = \frac{\hbar\mathbf{k}.\mathbf{p}}{m_0 \Delta E}$, where $\mathbf{p}$ is a typical matrix element of momentum between bands, and $\Delta E$ a typical interband separation. The order of magnitude of $S$ may be roughly estimated by assuming that $\mathbf{p}$ is of the order of $\hbar/a$, where $a$ is the lattice spacing, and that $\Delta E$ is of the order of $\hbar^2/ma^2$; then $S \sim a|\mathbf{k}|$. Thus, in general, the approximation is valid as long as $|\mathbf{k}|$ is much smaller than the width of the Brillouin zone which is $\sim 1/a$. However, in certain semiconductors, for example InSb, and in semimetals, the energy gap is so small that $S$ is abnormally large, so that the region of validity of the approximation is small.

The dependence of $m^*$ on the momentum matrix elements between bands provides a connection between $m^*$ and the optical properties, because the same matrix element governs the magnitude of the absorption due to interband transition. For instance, the value of the matrix element can be deduced from the effective mass (determined, say, by cyclotron resonance experiments) and used to predict a value for the optical absorption coefficient.

When an electric field is applied, $\mathbf{k}$ changes and may go outside the range in which the effective mass approximation is valid. If $\tau$ is the mean time between collisions, the change in $\mathbf{k}$ is of order $e\mathscr{E}\tau/\hbar$, and in the state reached by the electron the quantity $S$ which determines the validity of the approximation must still be small. For Ge, $S$ is still $<1$ for fields of thousands of V cm$^{-1}$, but for $n$-type InSb it reaches unity for only a few V cm$^{-1}$.

## 4.6 Motion in a magnetic field

The effective mass approximation can be applied in many cases to study the motion of an electron in a state near a band edge also when an external magnetic field $\mathbf{B}$ is applied. If $\mathbf{B}$ is in the direction of the $z$-axis, the vector potential $\mathbf{A}$ can be taken as

$$A_x = A_z = 0$$

$$A_y = Bx$$

With the notation

$$\mathbf{P} = \mathbf{p} - \frac{e\mathbf{A}}{c}$$

Schrödinger's equation is

$$\left[\frac{P^2}{2m_0} + V(\mathbf{r})\right]\Psi(\mathbf{r}) = E\Psi(\mathbf{r})$$

and, as before, this equation is replaced by an equivalent one in which the periodic potential $V(\mathbf{r})$ has disappeared and the free electron mass has been replaced by the effective mass at the band edge:

$$\frac{P^2}{2m^*} g_n(\mathbf{r}) = E'_n g_n(\mathbf{r}) \tag{4.52}$$

where $g_n(\mathbf{r})$ is an envelope function like (4.22) and

$$E'_n = E_n - E_n(0)$$

We consider the simplest case, in which $m^*$ is a scalar. Then, written in full, (4.52) is

$$-\frac{1}{2m^*}\left[\hbar^2\frac{\partial^2}{\partial x^2} + \hbar^2\frac{\partial^2}{\partial z^2} + \left(\frac{\hbar}{i}\frac{\partial}{\partial y} - \frac{eBx}{c}\right)^2\right]g_n = E'_n g_n \tag{4.53}$$

The operator on the l.h.s. does not contain the coordinates $y$ and $z$ explicitly, therefore $p_y$, $p_z$ commute with the Hamiltonian and are constants of the motion. Accordingly $g_n(\mathbf{r})$ has the form

$$g_n(\mathbf{r}) = e^{i(p_y y + p_z z)/\hbar} f(x) \tag{4.54}$$

and the eigenvalues of $p_y p_z$ can take any value from $-\infty$ to $+\infty$. $p_z$, in the direction of the magnetic field, remains unquantised; $p_y$, which appears in the combination $p_y - eBx/c$, will be discussed below. Substituting (4.54) into (4.53) we obtain the equation for $f(x)$

$$-\frac{\hbar^2}{2m^*}\frac{d^2f}{dx^2} + \frac{p_z^2 f}{2m^*} + \frac{p_y^2 f}{2m^*} - \frac{2eBp_y x}{2cm^*}f + \frac{e^2B^2x^2}{2m^*c^2}f = E'f \tag{4.55}$$

or, with the abbreviation

$$x_0 = \frac{cp_y}{eB} \tag{4.56}$$

$$\frac{d^2f}{dx^2} + \frac{2m^*}{\hbar^2}\left[E' - \frac{p_z^2}{2m^*} - \frac{e^2B^2}{2c^2m^*}(x - x_0)^2\right]f = 0 \tag{4.57}$$

This equation is the Schrödinger equation for a linear harmonic oscillator of angular frequency

$$\omega_c = \frac{eB}{cm^*} \tag{4.58}$$

oscillating about the point $x = x_0$. $\omega_c$ is called the cyclotron frequency and is numerically equal to $2.799\frac{m_0}{m^*} B$ MHz if $B$ is measured in gauss.

Hence $E' - p_z^2/2m^*$, which takes the place of the oscillator energy, must have the values

$$(N + \tfrac{1}{2})\hbar\omega_c = (N + \tfrac{1}{2})\frac{e\hbar B}{cm^*} \tag{4.59}$$

Thus the eigenvalues of (4.53) are

$$E_n - E_n(0) = (N + \tfrac{1}{2})\frac{e\hbar B}{cm_n^*} + \frac{\hbar^2 k_z^2}{2m_n^*} \tag{4.60}$$

where $k_z$ is the wave vector component in the direction of the magnetic field. $N$ can take the values 0, 1, 2, . . . and labels the so-called Landau levels. Thus the bottom of the band is shifted by $\dfrac{e\hbar B}{2cm^*}$ (about $4 \times 10^{-3}$ eV in InSb for a field of $10^4$ gauss) and the band splits near the edge into a succession of bands separated in energy by the constant quantity

$$\frac{e\hbar B}{cm^*} = \hbar\omega_c$$

The spacing is proportional to $B$ and thus the number of occupied Landau levels varies with $B$ and we may anticipate an oscillatory behaviour of physical properties as $B$ increases.

In the $k_x - k_y$ plane the electronic states, which are distributed uniformly in the absence of field, occupy now only certain circles with radii proportional to $(N + \tfrac{1}{2})^{\frac{1}{2}}$ (Fig. 4.2). They are not fixed at given points on the circle, but rotate round it with frequency $\omega_c$. But the levels can be classified by using the number $N$ of the circle on which they lie.

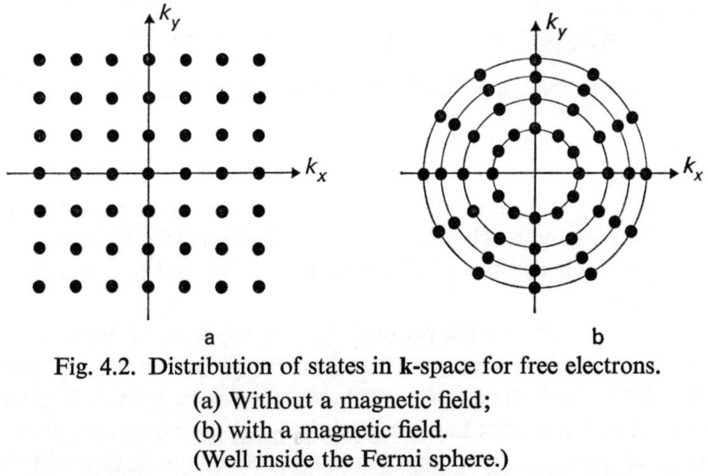


Fig. 4.2. Distribution of states in **k**-space for free electrons.
(a) Without a magnetic field;
(b) with a magnetic field.
(Well inside the Fermi sphere.)

In three dimensions, in the simplest case in which the Fermi surface is a sphere, at zero field the electrons occupy this sphere in momentum space. When the field is applied, the sphere splits up into a pile of cylindrical slices, each slice accomodating, in first approximation, all the electrons which in zero field were included in an interval $\Delta k_z$ of the order of $[(2N + 1)eB/c\hbar]^{\frac{1}{2}}$ (cf 4.60). The maximum electron energy is still the Fermi energy $E_F$, but the average energy depends on $B$, so that also the magnetic susceptibility $\chi = -d^2E_{av}/dB^2$ depends on $B$. As $B$ increases, the area of the cylinders expands and when the equatorial cylinder reaches a radius corresponding to $E_F$, or, more generally, when its area reaches the maximum cross-sectional area of the Fermi surface in planes perpendicular to the field, it empties of electrons, its role being taken over by the two adjacent cylinders. As $B$ is further increased, the process is repeated.

The change in occupation of the Landau levels as $B$ increases is illustrated in Fig. 4.3. In (a) the Fermi level, assumed to delimit a distribution dropping sharply to zero, is half way between two levels, so that the number of levels

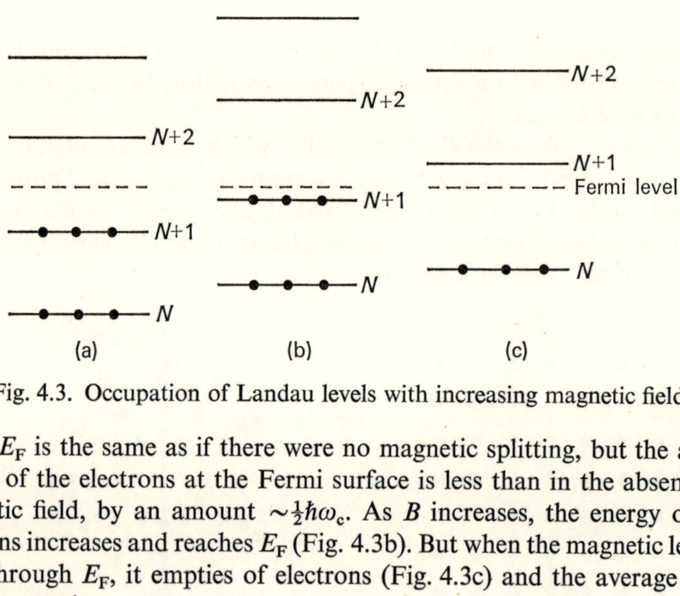


Fig. 4.3. Occupation of Landau levels with increasing magnetic field.

below $E_F$ is the same as if there were no magnetic splitting, but the average energy of the electrons at the Fermi surface is less than in the absence of a magnetic field, by an amount $\sim\frac{1}{2}\hbar\omega_c$. As $B$ increases, the energy of these electrons increases and reaches $E_F$ (Fig. 4.3b). But when the magnetic level has gone through $E_F$, it empties of electrons (Fig. 4.3c) and the average energy decreases again.

In fact, at temperature $T$ the probability of occupation of the various levels does not drop to zero abruptly at the Fermi energy, but gradually, within an interval $\sim kT$ ($k$ = Boltzmann's constant). However, provided the energy difference between adjacent Landau levels is $>kT$, their existence gives rise to oscillations of the magnetic susceptibility, and related quantities, with varying magnetic field (de Haas – van Alphen effect) of the form $\cos\left(\frac{a}{B} + b\right)$, where $a, b$ depend on the angle between the field and the crystal axes. From measurement of the period of these oscillations, one obtains the maximum cross-sectional area of the Fermi surface $A_0$, in planes perpendicular to the field. By repeating the experiment with $B$ at different angles to the crystal axes, $A_0$ is determined in all directions and then it is possible to reconstruct the Fermi surface itself, provided it has a centre of symmetry. All the effects connected with the quantisation of the electron orbits in a magnetic field provide in principle information on the Fermi surface.

Returning to the Landau levels, if spin is considered, the usual expression for the spin energy $2\mu_0 s_z B$ is just added up, and it is found that no spin splitting occurs, because the energy of the $n$ level with 'spin up' is equal to that of the $(n + 1)$ level with 'spin down'. $s_z$ is the spin quantum number and $\mu_0$ is the Bohr magneton corresponding to mass $m^*$,

$$\mu_0 = \frac{e\hbar}{2m^*c}$$

The un-normalised wave functions are, omitting the band index $n$,

$$g_N(\mathbf{r}) = e^{ip_y y/\hbar + ip_z z/\hbar}\, e^{-(x-x_0)^2/2L^2}\, H_N\left(\frac{x - x_0}{L}\right) \tag{4.61}$$

where $p_y = \hbar k_y$ may be considered as a quantum number labelling the degenerate functions belonging to each eigenvalue $(N + \frac{1}{2})\hbar\omega_c$; $L$ is a characteristic length: $L^2 = \frac{\hbar c}{eB}$, $L = \frac{3{\cdot}4\ 10^{-4}}{B^{\frac{1}{2}}}$ cm, when $B$ is measured in gauss; and $H_N$ is the Hermite polynomial of order $N$. In classical mechanics, the motion in the $x$–$y$ plane, with angular frequency $\omega_c$, is circular; $x_0$ corresponds to the classical $x$-coordinate of the centre of this circle. It may be proved that the $y_0$-coordinate is $L^2 k_y - y$, which is a constant of the motion, like $x_0$. However, $x_0$ and $y_0$ cannot be measured simultaneously. (For more details see, for instance, Landau and Lifschitz, 1958, p. 471).

The above treatment holds provided the function $f_N(x)$ is slowly varying. For this is it necessary that $L$ should be much larger than the lattice spacing. Except in exceedingly high fields, this is true for $N = 0$. As $N$ increases, the Hermite polynomial superimposes a more rapid variation and the effective mass approximation becomes less valid. The value of $N$ at which it breaks down varies from material to material.

When the inverse mass is a tensor, the essential features of the theory are the same, but the treatment may become involved.

## 4.7 Cyclotron resonance

Often the motion in a magnetic field can be treated in a semi-classical way. Classically, under the influence of a steady uniform magnetic field **B**, an electron in a crystal moves according to the law

$$\hbar\dot{\mathbf{k}} = \frac{1}{c}\mathbf{w} \times \mathbf{B} \tag{4.62}$$

Thus **k** changes only in planes perpendicular to the field, and the orbits in **k**-space are the intersections of these planes with the constant energy surfaces. In true space the electrons describe helices, the axis of the helices being parallel to **B**.

To find the period with which the orbit is described, consider two orbits in

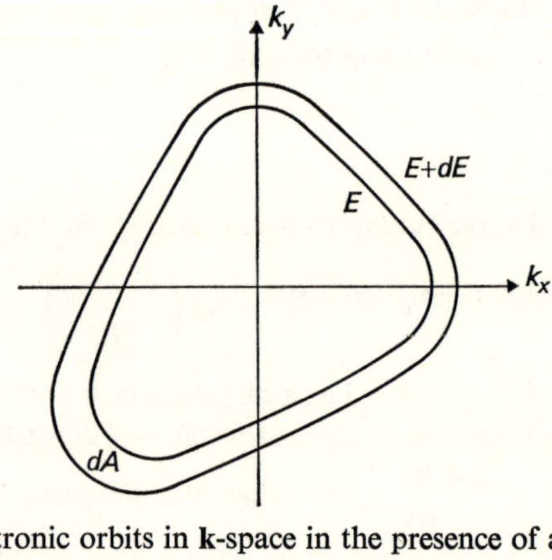


Fig. 4.4. Electronic orbits in **k**-space in the presence of a magnetic field.

**k**-space, in a plane perpendicular to **B** (Fig. 4.4) corresponding to energies $E$ and $E + dE$. The increment $d\mathbf{k}$ in going from one orbit to the other is

$$d\mathbf{k} = \frac{dE}{|\text{grad}_{\mathbf{k}}\, E|_{\perp \mathbf{B}}} = \frac{dE}{\hbar \mathbf{w}_{\perp B}} \tag{4.63}$$

The area of **k**-space between the two orbits is

$$dA = \frac{d}{\hbar}\oint \frac{dk}{\mathbf{w}_{\perp \mathbf{B}}} \tag{4.64}$$

Writing $d\mathbf{k} = \dot{\mathbf{k}}\, dt$, we have

$$T = \oint \frac{dk}{|\dot{\mathbf{k}}|}$$

and then, from (4.62), (4.64),

$$T = \frac{c\hbar}{eB} \oint \frac{dk}{\mathbf{w}_{\perp \mathbf{B}}} = \frac{c\hbar^2}{eB} \frac{dA}{dE} \tag{4.65}$$

so that the cyclotron frequency is

$$\omega_c = \frac{2\pi}{T} = \frac{2\pi e B}{c\hbar^2} \bigg/ \frac{dA}{dE} \tag{4.66}$$

$\omega_c$ is thus different for different orbits. Comparing (4.66) and (4.58), we see that the 'cyclotron effective mass' for the orbit is

$$m_c^* = \frac{\hbar^2}{2\pi} \frac{dA}{dE} \tag{4.67}$$

In semiconductors, where the charge carriers are concentrated at one, or a few equivalent points in the Brillouin zone, all the orbits have the same frequency for a given field direction. In metals the detailed shape of the Fermi surface must be taken into consideration.

One of the most useful experimental method for the determination of effective masses is cyclotron resonance. In this experiment, a radio frequency is applied, besides the constant magnetic field, and the electron absorbs energy when the frequency of this field equals the cyclotron frequency. The experiment must be performed at a low temperature in order that the electron mean free path be long enough to allow it to describe a few orbits before being scattered. According to quantum theory, we would say that an impressed radiation field of angular frequency $\omega_c$ induces transitions which change the value of the quantum number $N$ of the Landau levels. But for an oscillator we have the selection rule that $N$ can change only by $\pm 1$. Hence the emitted or absorbed photon has energy $\hbar\omega_c$, in agreement with the classical interpretation based on the idea of resonance.

As an example, consider the classical treatment of a cyclotron resonance experiment aimed at investigating the conduction band edge in silicon or germanium. Here one wishes to check the model that, near the edge, the equienergy surfaces consist of rotation ellipsoids at equivalent positions in **k**-space, and to determine the shape of these ellipsoids, that is the principal effective masses. Consider one rotation ellipsoid and take the $z$-axis along its symmetry axis; measure **k** from the centre of the ellipsoid. Near the edge then

$$E = E_0 + \frac{\hbar^2}{2m_t}(k_x^2 + k_y^2) + \frac{\hbar^2}{2m_l} k_z^2 \tag{4.68}$$

$m_l$ is called the longitudinal mass, and $m_t$ the transverse mass. The equations of motion are

$$\mathbf{w} = \frac{1}{\hbar}\,\mathrm{grad}_{\mathbf{k}}\, E(\mathbf{k})$$

$$\hbar\frac{d\mathbf{k}}{dt} = \frac{e}{c}\mathbf{w} \times \mathbf{B} \tag{4.69}$$

For the spheroidal surface under consideration

$$w_x = \frac{\hbar k_x}{m_t}, \qquad w_y = \frac{\hbar k_y}{m_t}, \qquad w_z = \frac{\hbar k_z}{m_l}$$

Take the magnetic field in the $x$–$z$ plane, making angle $\theta$ with the $z$-axis:

$$B_x = B\sin\theta, \qquad B_y = 0, \qquad B_z = B\cos\theta$$

Substituting into (4.69) and assuming that $\mathbf{k}$ varies with time through a factor $e^{i\omega t}$, where $\omega$ is the cyclotron frequency, we have

$$i\omega k_x = \frac{e}{cm_t}k_y B\cos\theta$$

$$i\omega k_y = \frac{e}{cm_l}k_z B\sin\theta - \frac{e}{cm_t}k_x B\cos\theta \tag{4.70}$$

$$i\omega k_z = \frac{e}{cm_t}k_y B\sin\theta$$

These are linear homogeneous equations in $k_x$, $k_y$, $k_z$: the determinant of the coefficients must vanish. Let

$$\frac{eB}{cm_t} = \omega_t, \qquad \frac{eB}{cm_l} = \omega_l$$

then

$$\begin{vmatrix} i\omega & -\omega_t\cos\theta & 0 \\ \omega_t\cos\theta & i\omega & -\omega_l\sin\theta \\ 0 & \omega_t\sin\theta & i\omega \end{vmatrix} = 0 \tag{4.71}$$

or

$$\omega^2 = \omega_l\omega_t\sin^2\theta + \omega_t^2\cos^2\theta$$

Writing $\omega$ as $eB/cm^*$, the effective mass determining the cyclotron frequency when the static magnetic field makes an angle $\theta$ with the longitudinal axis of the energy surface is

$$\left(\frac{1}{m^*(\theta)}\right)^2 = \frac{\cos^2\theta}{m_t^2} + \frac{\sin^2\theta}{m_l m_t} \tag{4.72}$$

Thus, by measuring $m^*(\theta)$ at various angles to fixed axes, $m_l$ and $m_t$ can be determined, and the direction of the axis of the ellipsoid. Experiments have confirmed the validity of the model, showing that in silicon the ellipsoids are along the [1,0,0]-directions, and in germanium at the boundary of the Brillouin zone along the [1,1,1]-directions. In interpreting the experimental results one must of course take into account the existence of several equivalent ellipsoids. Fortunately, each valley can be treated independently, as the resonance, except for particular directions of **B**, occurs at a different frequency for each.

By repeating the experiment with different orientations of **B** with respect to the crystal axes, information on the band edge symmetry can be obtained.

## 4.8 The concept of positive hole

If an energy band is completely occupied, except for the state with vector **k**, one says that a hole with such vector is present. A hole moves in a field as if it had an effective mass exactly opposite to the effective mass of the electron with the same crystal momentum

$$\left(\frac{1}{m_h^*(\mathbf{k})}\right) = -\left(\frac{1}{m_e^*(\mathbf{k})}\right) \tag{4.73}$$

Holes occur near the upper edges of bands, where the electron effective mass is usually negative (exceptions due to the peculiar shapes of the energy bands are unimportant), and thus their effective mass is positive. They also have a positive charge.

The proof of the above statements is based on the fact that the total current carried by the electrons of a completely full band vanishes (Section 4.4) and that the average acceleration of these electrons is also zero. To see this, consider a one-dimensional case when the Brillouin zone extends from $-k_0$ to $+k_0$. The average acceleration $\bar{a}$ is proportional to the integral

$$\bar{a} = \int_{-k_0}^{k_0} \frac{F}{m^*}\, dk$$

Now $F$ is independent of $k$, so that, using (4.35),

$$\bar{a} = \frac{F}{\hbar^2}\left[\frac{dE}{dk}\right]_{-k_0}^{+k_0}$$

But $dE/dk$, proportional to the velocity, vanishes at both limits, thus $a = 0$. The same type of proof can be given in three dimensions.

If now the band is full, except for one state with wave vector $\mathbf{k}'$, the acceleration is given by the same integral as before, less the acceleration of the electrons with wave vector $\mathbf{k}'$:

$$a = 0 - \frac{1}{m^*(\mathbf{k}')}\mathbf{F}$$

that is, the entire electron assembly is accelerated as though it consisted of only one particle (hole) with wave vector $\mathbf{k}'$ and with an effective mass tensor of opposite sign to that pertaining to the missing electron. As remarked, the commonest case is a missing electron near the top of a band, where $m_e^*$ is negative, so that $m_h^*$ is positive.

As for the current, we separate the total current density $\mathbf{j}$, which is zero, into a part $\mathbf{j}_1$ due to the electron with wave vector $\mathbf{k}'$, and a part $\mathbf{j}_2$ due to all the other electrons,

$$\mathbf{j}_1 + \mathbf{j}_2 = 0$$

Now

$$\mathbf{j}_1 = -e\mathbf{w}(\mathbf{k}')$$

and thus

$$\mathbf{j}_2 = +e\mathbf{w}(\mathbf{k}')$$

If the electron with wave vector $\mathbf{k}'$ is removed, the current is $+e\mathbf{w}(\mathbf{k}')$. Thus the lack of an electron from an otherwise full band is equivalent to a positive particle (positive hole, or hole for short) with usually positive mass in all directions.

In practice it is convenient to measure the energy of holes in opposite direction to that of electrons,

$$E_h(\mathbf{k}) = -E_e(\mathbf{k}) + \text{constant} \tag{4.74}$$

With this convention the definition (4.33) of the inverse mass tensor applies to both kinds of carriers, and the band edge is always the point of lowest energy. For a hole with scalar effective mass $m_h^*$, the relation between velocity and crystal momentum is

$$\mathbf{w} = -\frac{\hbar}{m_h^*}k$$

If an electron-hole pair is created by absorption of a photon of energy $E$ (equal to, or slightly larger than the energy gap), that is if an electron is excited from the valence to the conduction band, the electron and hole have practically the same value of $\mathbf{k}$, because of conservation of momentum, since the momentum $E/c$ of a photon is small. Thus their velocities are in opposite directions.

Changing the sign of the charge, holes contribute in the same way as electrons to the electric, magnetic, thermal, etc. properties of the crystal. The spin $\frac{1}{2}$ may be attributed to the hole. Holes obey equations of motion perfectly analogous to those for electrons. One recognises the similarity between the concept of a hole and that of a positron, at least in the original Dirac formulation, since both can be interpreted as the absence of a negative charge from a state, in an otherwise full distribution, in which its mass would be negative.

Although the proof we have sketched of (4.73) was based on the periodic structure of crystals, it can be shown that the concept is valid much more generally; for instance conduction by holes arises in liquids and amorphous solids.

# 5

# Band calculation

## 5.1 Free electron approximation

The free electron approximation is based on the early idea of Drude and Lorentz that electrons in metals are not subject to any force, in the absence of external fields. In band theory this idea is coupled to the assumption of periodicity, so that we speak of an empty lattice, in which the constant potential $V$ is supposed to be the limit of a periodic potential; the value of $V$ may always be taken as zero. The electrons are represented by plane waves

$$\Psi_{\mathbf{k},n}(\mathbf{r}) = \frac{1}{\sqrt{\Omega}} e^{i(\mathbf{k}+\mathbf{K}_n)\cdot\mathbf{r}} \tag{5.1}$$

where $\mathbf{K}_n$ is any vector of the reciprocal lattice. For any value of $\mathbf{k}$ there is an infinite number of states, each denoted by a reciprocal lattice vector $\mathbf{K}_n$ playing the role of the band index, whose energies are given by (2.44)

$$E_n(\mathbf{k}) = \frac{\hbar^2}{2m_0}|\mathbf{k} + \mathbf{K}_n|^2 \tag{5.2}$$

Using (5.2) the 'free electron band structure' is constructed without any difficulty. Despite the drastic approximation used, it is of some practical value. Its most important application is for predicting the shape of the Fermi surface of metals. To illustrate the procedure, consider as an example the face-centred cubic lattice. The Brillouin zone is the truncated octahedron of Fig. 2.8, shown again in Fig. 5.1. The centre point $\mathbf{k} = 0$ is conventionally denoted by $\Gamma$; other symmetry points are shown: the points $X$ are along the [1,0,0] and equivalent directions, at the centre of a square face. Each point $X$ is equivalent to the opposite point $X$, since they are separated by a vector of the reciprocal lattice; there are thus three non-equivalent $X$-points. Points along the same directions, from $\Gamma$ to $X$ are denoted by $\Delta$; the points $L$ are along the [1,1,1] and equivalent directions, at the centre of one of the hexagonal faces: there are four non-equivalent $L$-points; points on the same directions from

$\Gamma$ to $L$ are denoted by $\Lambda$; $K$ is any point along the [1,1,0] and equivalent directions; $W$ is the end point of the [0,2,1] and equivalent directions; the 24 $W$-points are equivalent in group of four, so that there are six non-equivalent ones.

Assuming for convenience that the side of the fundamental cube in the

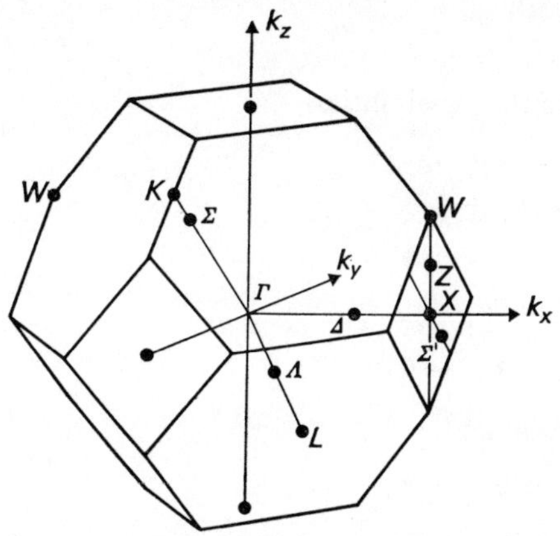


Fig. 5.1. Brillouin zone for the face-centred cubic lattice. Points and lines of symmetry are indicated.

direct lattice is $2\pi$, then in the Brillouin zone the distance $\Gamma$–$X$ is one. Written in terms of Cartesian components, equation (5.2) is (atomic units)

$$E(k_x, k_y, k_z, n) = (k_x + K_{nx})^2 + (k_y + K_{ny})^2 + (k_z + K_{nz})^2 \quad (5.3)$$

The constant energy surfaces are spheres centred at all the reciprocal lattice points. The shortest vectors of the reciprocal lattice (body-centred) are

$\mathbf{K}_0 = 0$;
$\mathbf{K}_1 = (\pm 1, \pm 1, \pm 1)$: eight vectors of length $\sqrt{3}$;
$\mathbf{K}_2 = (\pm 2, 0, 0)$: six vectors of length 2;
$\mathbf{K}_3 = (\pm 2, \pm 2, 0)$: twelve vectors of length $\sqrt{8}$;
$\mathbf{K}_4 = (\pm 3, \pm 1, \pm 1)$: twenty four vectors of length $\sqrt{11}$;
$\mathbf{K}_5 = (\pm 2, \pm 2, \pm 2)$: eight vectors of length $\sqrt{12}$, etc.

Thus the successive energy levels at $\mathbf{k} = 0$ are

$$E_0 = 0, \quad E_1 = 3, \quad E_2 = 4, \quad E_3 = 8, \ \ldots$$

$E_1$ is eight times degenerate, $E_2$ six times, etc.

Consider now one of the $\Delta$-directions, for instance [0,1,0]. Along this direction

$$k_x = k_z = 0, \qquad 0 \leqq k_y \leqq 1$$

it is not necessary to consider negative values of $k_y$ because

$$E(\mathbf{k}) = E(-\mathbf{k})$$

Starting with $\mathbf{K}_0$, equation (5.3) gives the non-degenerate band

$$E_0 = k_y^2$$

Taking $\mathbf{K}_1$, we have the possibilities

$$\begin{aligned} k_x + \mathbf{K}_x &= \pm 1, \\ k_y + \mathbf{K}_y &= k_y \pm 1, \\ k_z + \mathbf{K}_z &= \pm 1 \end{aligned}$$

with two possible bands:

$$E_1^{(1)} = 2 + (k_y - 1)^2$$

four times degenerate, and

$$E_1^{(2)} = 2 + (k_y + 1)^2$$

also four times degenerate. At $\mathbf{k} = 0$, both bands start from the eight times degenerate level $E_1 = 3$; for $E_1^{(1)}$ the energy decreases as $k_y$ increases; for $E_1^{(2)}$ the energy increases. Taking now $\mathbf{K}_2$, there are the possibilities

$$\begin{aligned} &k_x + K_x = \pm 2, \quad && k_y + K_y = k_y, \quad && k_z + K_z = 0, \\ &k_x + K_x = 0, && k_y + K_y = k_y \pm 2, && k_z + K_z = 0, \\ &k_x + K_x = 0, && k_y + K_y = k_y, && k_z + K_z = \pm 2 \end{aligned}$$

and thus three possible bands

$$\begin{aligned} E_2^{(1)} &= 4 + k_y^2, && \text{four times degenerate,} \\ E_2^{(2)} &= (k_y - 2)^2, && \text{non-degenerate,} \\ E_2^{(3)} &= (k_y + 2)^2, && \text{non-degenerate} \end{aligned}$$

These bands are shown in Fig. 5.2a. The bands along the [1,1,0], [1,2,0], [1,1,1] are calculated in the same way and are shown, respectively, in Figs. 5.2b, c, d. $K_x$, $K_y$, $K_z$ are indicated on each curve and (in brackets) the degeneracy, The degeneracy of the states at $\Gamma$, $W$, $K$, $X$, $L$ is also indicated. Fig. 5.3 illustrates the bands along a direction of no symmetry, the [5,6,7]-direction: at points of no symmetry, all the bands are in general non-degenerate, though there may be accidental degeneracies. The position of the Fermi level, in atomic units,

$$E_F = \left(\frac{3Z}{2\pi}\right)^{\frac{2}{3}}$$

is also indicated in Figs. 5.2, 5.3, as a function of the number $Z$ of electrons per unit cell ($E_F$ is given generally by $\hbar^2(3\pi^2 n)^{2/3}/2m_0$ – see the next Section – where $n$ is the number of free electrons per unit volume. The number $Z$ of electrons in the unit cell of our face-centred cubic lattice is $\frac{1}{4}(2\pi)^3 n$).

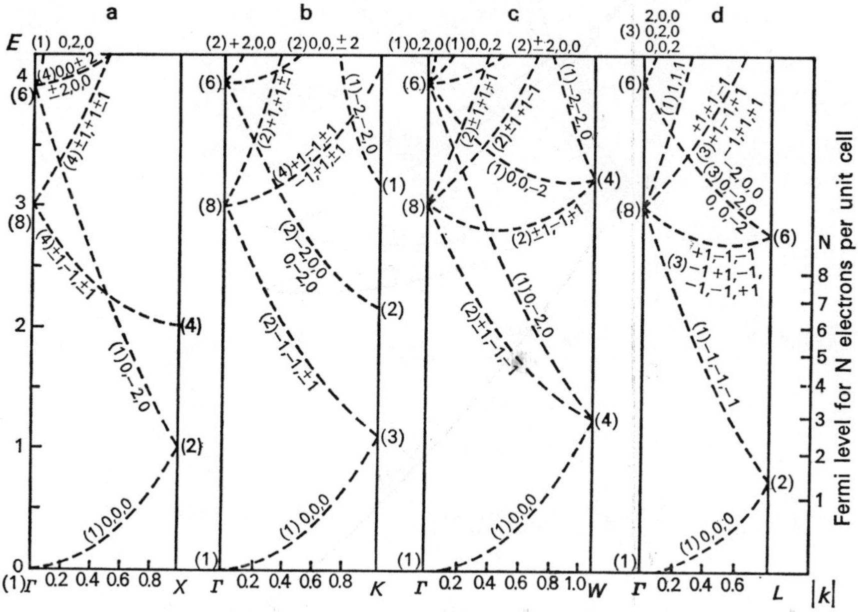


Fig. 5.2. Energy bands for the face-centred cubic structure in the free electron approximation, along symmetry directions.

We shall see later (Section 5.3) that bands which cross in this approximation, for example bands $(\pm1,-1,\pm1)$ and $(0,-2,0)$ in Fig. 5.2a, do not generally cross once the periodic field is taken into consideration. The exceptions, as already remarked (Section 2.10), occur along symmetry lines, where the possibility arises of bands with different symmetry (Section 6.1) which can cross. Every time that the free electron scheme indicates a forbidden crossing of bands, the approximation becomes poor and the true band structure is essentially different.

But often this simple picture allows us to make qualitative predictions on the band structure of metals. For instance, it is clear that in divalent metals there may be electrons in the second band at some **k**-points, and holes left in the first band at other $k$-points. This band overlapping explains why alkaline-earth elements are metals, while one might have thought that the two electrons

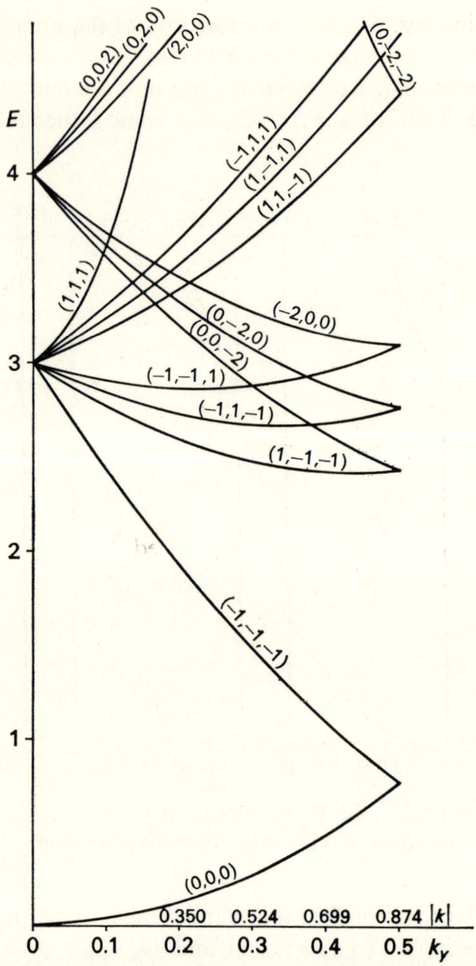


Fig. 5.3. Energy bands for the face-centred cubic structure in the free electron approximation, along a direction of no symmetry. All the bands are non-degenerate.

would fill completely one band, giving rise to an insulator. It explains also why in these metals one may observe positive hole conduction.

The free electron model can give some indication also on the band structure of semiconductors. For a detailed discussion see Herman (1958).

## 5.2 Construction of the Fermi surface

In the free electron approximation the surfaces of constant energy are spheres in **k**-space. The allowed values of **k** are distributed with uniform density

$V/4\pi^3$ in this space, for a crystal of volume $V$, taking spin into account. If there are $Z$ electrons per unit cell and $N$ cells per unit volume, according to the Pauli principle, at 0 K all the states up to $k_F$ are filled, where

$$\frac{4\pi}{3}k_F^3\frac{1}{4\pi^3} = ZN$$

$$k_F = (3\pi^2 ZN)^{\frac{1}{3}} \tag{5.4}$$

The sphere of radius $k_F$ is called the Fermi sphere. However, when the periodic nature of the structure is taken into account, and the reduced zone scheme is used, the Fermi surface is not generally a sphere. The Fermi energy, or Fermi level, is the energy corresponding to the top of the distribution

$$E_F = \frac{\hbar^2 k_F^2}{2m_0} = \frac{\hbar^2}{2m_0}(3\pi^2 n)^{\frac{2}{3}} \tag{5.5}$$

where $n = ZN$ is the number of free electrons per unit volume.

The average energy of the electrons is obtained by integration over the Fermi sphere

$$\bar{E} = \frac{\int_0^{k_F} \frac{\hbar^2 k^2}{2m_0} 4\pi k^2 dk}{\frac{4}{3}\pi k_F^3} = \frac{3}{5}\frac{\hbar^2 k^2}{2m_0} = \frac{3}{5}E_F$$

The total electron energy per unit volume, called zero point energy, is

$$U_0 = n\bar{E} = \frac{3\hbar^2}{10m_0}(3\pi^2)^{2/3}n^{5/3}.$$

As for ideal gases, the pressure (zero point pressure) of the electron gas is $p_0 = \frac{2}{3}U_0$ and the equation of state is $p_0 V = \frac{2}{3}nV\bar{E}$. In complete contrast to the classical gas, the average energy is independent of temperature. The electron gas is then said to be completely degenerate. As the temperature increases, a dependence on the temperature appears, the degeneracy is incomplete. At very high temperatures ($T > E_F/k$) the classical laws are obeyed. At normal temperatures the degeneracy is almost complete, so that the electronic contribution to the specific heat of the metal, $dU/dT$, is negligible. However it can be proved that this small contribution is proportional to $T$, while, at low temperature, the lattice contribution is proportional to $T^3$; thus, at extremely low temperatures, the specific heat of a non-superconducting metal is determined by the electrons.

The radius of the Fermi sphere is of the order of the linear dimensions of the Brillouin zone and of the radius of the Debye sphere; the wavelength of an electron at the Fermi surface is of the order of the interatomic spacing. Fermi energies are of the order of a few eV. $E_F/k$ is of the order of $10^4 - 10^5$ K. $p_0$ is of the order of $10^5$ atmospheres; $U_0$ is of the order of $10^4$ joule cm$^{-3}$.

From (2.53), the density of levels at the Fermi surface is $m_0 k_F/\pi^2\hbar^2$ and using (5.4),

$$\frac{dn_s}{dE}(E_F) = \frac{m_0}{\pi^2\hbar^2}(3\pi^2 n)^{\frac{1}{3}} \tag{5.6}$$

(half this number for electrons with one spin direction). Multiplying (5.5) by (5.6) we have

$$\frac{dn_s}{dE}(E_F) = \frac{3n}{2E_F} \tag{5.7}$$

For most metals the order of magnitude of $\frac{dn_s}{dE}(E_F)$ is $1 - 10$ states per eV per atom. Formula (5.7) was quoted in Section 4.2 in connection with the screen-

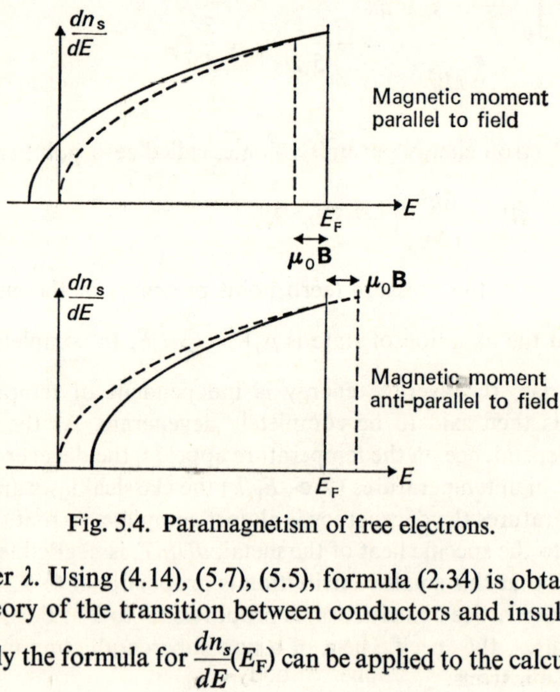


Fig. 5.4. Paramagnetism of free electrons.

ing parameter $\lambda$. Using (4.14), (5.7), (5.5), formula (2.34) is obtained, used in the Mott theory of the transition between conductors and insulators.

Incidentally the formula for $\frac{dn_s}{dE}(E_F)$ can be applied to the calculation of the paramagnetic susceptibility of a gas of free electrons. When there is no external magnetic field, all states up to the Fermi level are each occupied by two electrons with opposite spins. We can draw two density of states curves occupied up to the same energy, one for electrons with positive spin and one for electrons with negative spin. When a magnetic field **B** is applied, the energy of the electrons with magnetic moment $\mu_0$ parallel to the field is decreased by $\mu_0 B$ ($\mu_0$ = Bohr magneton = $e\hbar/2m_0c$) and the energy of the electrons with magnetic moment antiparallel to **B** is increased by $\mu_0 B$. The two density of states curves become displaced relative to each other by the energy $2\mu_0 B$, as shown in Fig. 5.4. This means that there are now more electrons with magnetic moment parallel to **B** than electrons with magnetic moment antiparallel to **B**. Since, even for fields of several thousand gauss, $\mu_0 B$ is only $\sim 10^{-5}$ eV, it can be assumed that the density of states for the electrons whose magnetic moment has turned over is $N(E_F)$. The difference in question is then $2\mu_0 BN(E_F)$. The magnetic moment per unit volume is

$$2\mu_0^2 BN(E_F) = 2\mu_0^2 B \frac{m_0}{2\pi^2\hbar^2}(3\pi^2 n)^{\frac{1}{3}}$$

and the paramagnetic susceptibility is $(m_0\mu_0{}^2/\pi^2\hbar^2)(3\pi^2 n)^{\frac{1}{3}}$ (temperature independent).

Note that the total susceptibility is made up of this spin paraagnetismm plus diamagnetic contributions from the ion cores and from the free electrons, due to the effect of **B** upon the orbital motion. The diamagnetic susceptibility of free electrons can be proved to be $-1/3$ the paramagnetic susceptibility. These values are found to be of the right order of magnitude for simple metals where the band structure does not affect the free electron formula. However experimental determinations are difficult because the paramagnetism is very small and any ferromagnetic impurity has a large effect on the measured values.

The Brillouin zone structure must now be taken into consideration. Consider a face-centred cubic metal, with $Z$ electrons per unit cell, then

$$k_F = \frac{1}{d}(12\pi^2 Z)^{\frac{1}{3}}$$

where $d$ is the side of the elementary cube. The Brillouin zone is the truncated octahedron of Fig. 5.1 where the distance of the hexagonal faces from the origin is $k_B = \sqrt{3}\pi/d$. Hence $k_F/k_B = 0{\cdot}903Z^{\frac{1}{3}}$. For alkali metals ($Z = 1$) the Fermi surface is entirely inside the Brillouin zone and the above considerations hold. But already for $Z = 2$, $k_F/k_B = 1{\cdot}14$. Apart from small changes in the numerical values, the same applies to body-centred metals.

When the Fermi surface cuts the Brillouin zone boundary, we must remember that the constant energy surfaces cut perpendicularly the zone faces – though in some cases this is true only along lines of symmetry – and thus we must expect the Fermi sphere to be much distorted near zone boundaries. Some distortion already occurs in alkali metals. Fig. 5.5 shows schematically

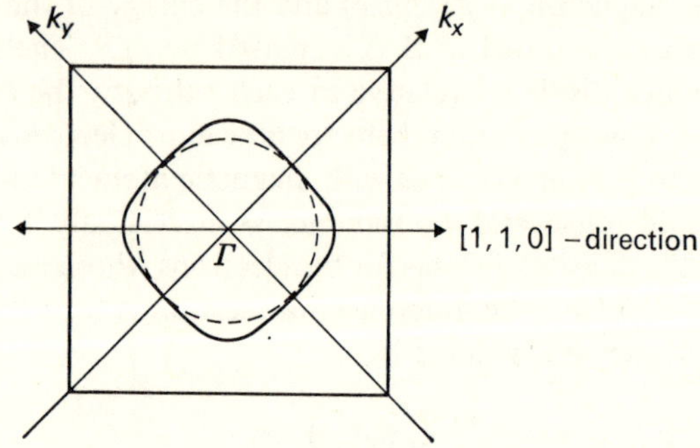


Fig. 5.5. Distortion of Fermi surface near a zone boundary.

the cross section of the Fermi surface of an alkali metal (b.c.c.) in a plane perpendicular to one of the [1,0,0]-directions. The dotted line is a section of the undistorted sphere. The full line indicates how it may get distorted. In alkali metals the distortion is small, but in the noble metals, copper, gold and silver (f.c.c.), the Fermi surface does cut the hexagonal faces of the Brillouin zone (Fig. 5.6). In the repeated zone scheme the Fermi surface be-

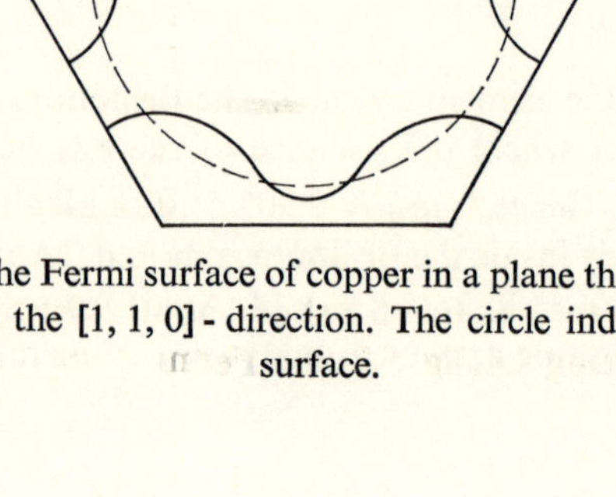


Fig. 5.6. Section of the Fermi surface of copper in a plane through the centre of the BZ perpendicular to the [1, 1, 0] - direction. The circle indicates the undistorted surface.

comes multiply connected and this has important physical consequences, mentioned in Section 4.4.

Using the reduced zone scheme, the portions of the Fermi surface outside the Brillouin zone must be displaced by a suitable reciprocal lattice vector in the usual way, as indicated schematically in Fig. 5.7. In this way, the Fermi

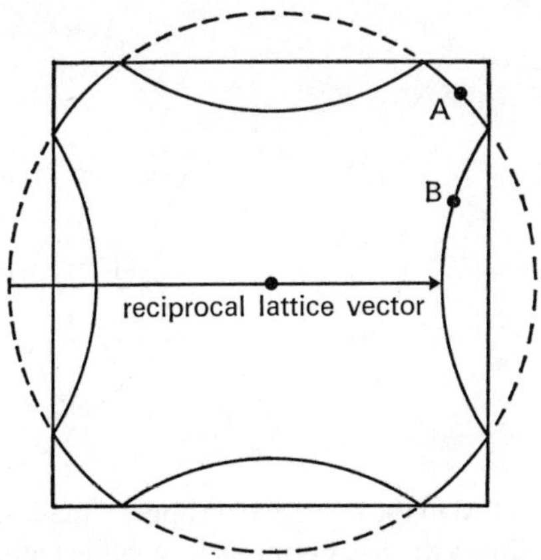


Fig. 5.7. Fermi surface in the reduced zone scheme (schematic).

surface is reflected inwards from the boundary and it is clear that the reflected surface is part of the Fermi sphere centred at one of the neighbouring reciprocal lattice points. This suggests a simple construction of the Fermi surface in the free electron approximation. Draw about each reciprocal lattice point a sphere whose radius corresponds to the number of free electrons present. The volume of the sphere is equal to the volume of the Brillouin zone if there are two free electrons per Bravais cell. The spheres intersect when there is more than one electron per atom. The intersections (Fig. 5.7) occur only on the surface of the (repeated) Brillouin zone. The Fermi surface changes spheres at these intersections (from A to B in Fig. 5.7).

Considering any point **k**, if it is contained within a single sphere, a free electron state at that point in the first band is occupied. If it lies within two spheres, states belonging to the first and second bands are occupied, and so on. If **k** lies outside all spheres, it is unoccupied. Thus the different parts of the Fermi surface separate regions of **k**-space occupied by different numbers of electrons. In Fig. 5.7 curve A is part of what may be called the 0–1 Fermi surface, as it separates an unoccupied region of **k**-space from a singly occupied region, while B is part of the 1–2 Fermi surface.

Considering actual metals, Fig. 5.8 shows as an example the Fermi surface of aluminium. Referring to Fig. 5.2, the Fermi energy, for $Z = 3$, is $E_F =$

1·2707. In this model then the first zone is completely occupied. It has been found that this is indeed true, though no firm conclusion could be reached from the free electron approximation since the highest energy in the first zone, $E = 1{\cdot}2247$ at the point $W$, is so near $E_F$. At practically all **k**-points the Fermi surface is in the second zone, but there are pockets of electrons in the third

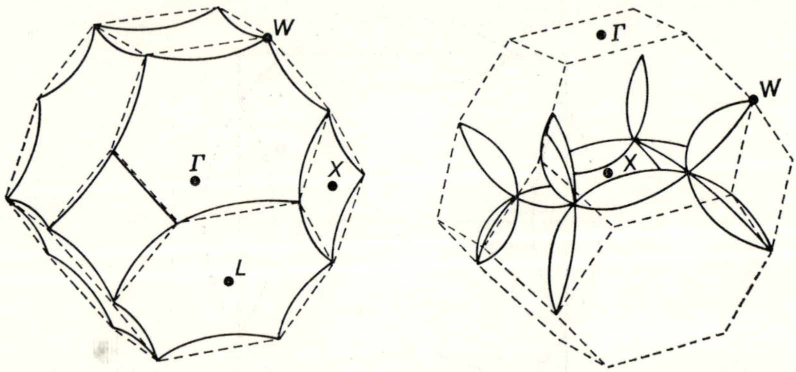


Fig. 5.8. Fermi surface of Al in the free electron approximation. The first band is full. The left-hand drawing shows the 1–2 Fermi surface in the reduced zone scheme. The right-hand drawing shows the 2–3 FS. In this last drawing the centre of the BZ has been taken at $X$, instead of $\Gamma$, in order to show the joining up of the pockets of electrons around the various corners. (After Harrison, 1966)

zone, around $W$. That is, at these points there are at least three allowed values of the energy below $E_F$. There could be three, or four, or five, depending on how the bands, degenerate at $W$ in this model, split in the actual

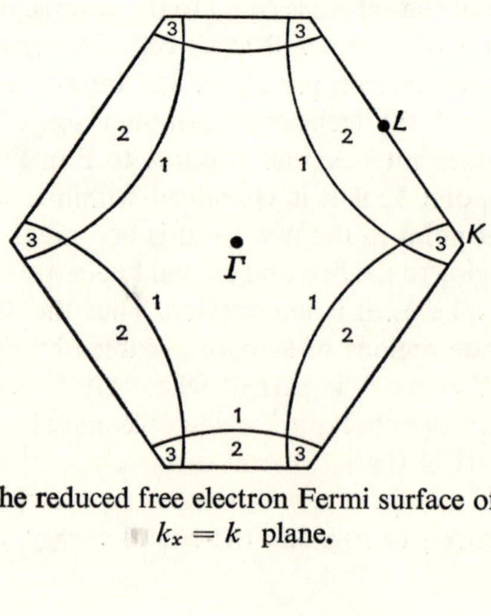


Fig. 5.9. Section of the reduced free electron Fermi surface of aluminium with the $k_x = k$ plane.

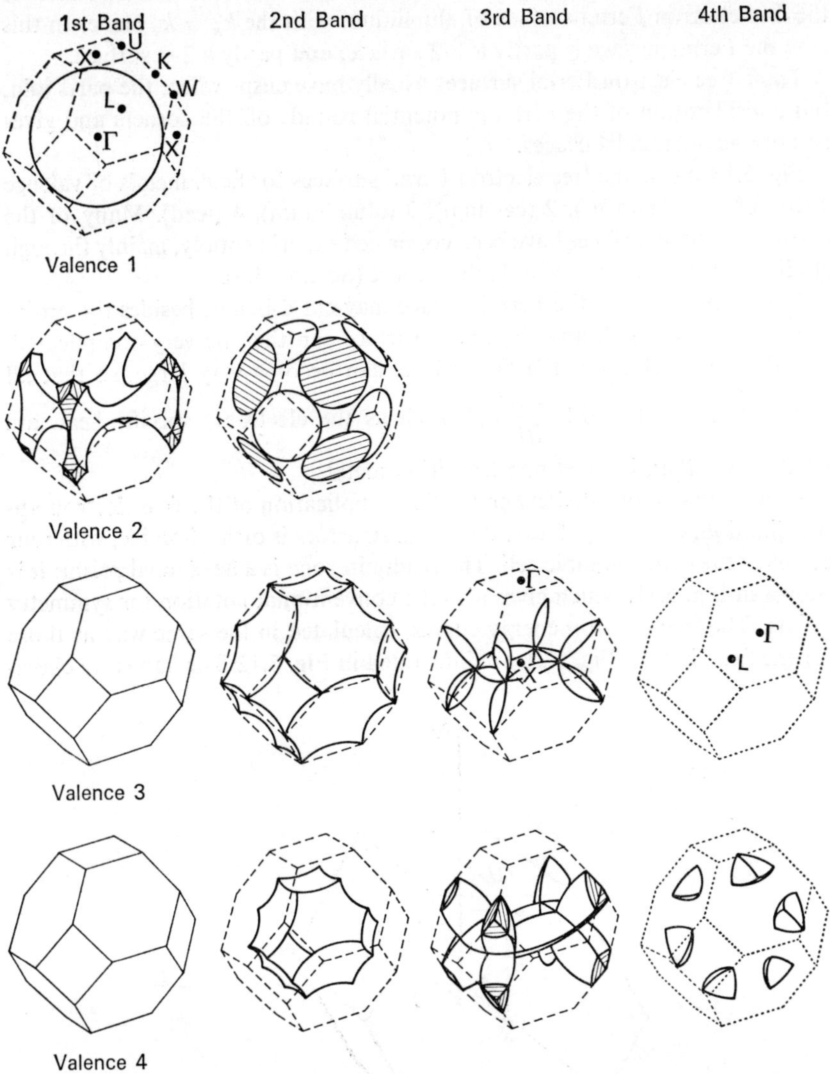


Fig. 5.10. Free electron Fermi surfaces for f.c.c. metals of valence 1, 2, 3 and 4. Regions with convex surfaces are occupied, regions with concave surfaces are unoccupied. The first band zone for valence 2, and the third band zones are centred at X; the fourth band zones are centred at L. (After Harrison, 1966)

crystal. This splitting will be discussed in Section 6.2. There seem in fact to be small pockets of electrons in the fourth zone. Fig. 5.9 shows the intersection of the free electron Fermi surface of aluminium with the $k_x = k_y$ plane. In this case the Fermi surface is partly a 1–2 surface, and partly a 2–3 surface.

These free electron Fermi surfaces usually have cusps where the parts join, but consideration of the periodic potential rounds off the corners and gives smooth geometrical surfaces.

Fig. 5.10 shows the free electron Fermi surfaces for f.c.c. metals of valence 1 (for example copper), 2 (calcium), 3 (aluminium), 4 (lead). Many of the features of these surfaces have been confirmed experimentally, mainly through studies of the de Haas – van Alphen effect (Section 4.6).

In transition metals the Fermi surface may cut d-bands, besides the ordinary conduction band, and the Fermi surface can then be very complicated. The density of states at the Fermi level can become very large, so that all quantities proportional to $\frac{dn_s}{dE}(E_F)$, such as the electronic specific heat, are much larger than in most non-transition metals.

As a more complicated example of the application of the free electron approximation, consider gallium. Its crystal structure is orthorhombic, with four atoms in the orthorhombic cell. The Brillouin zone is a hexagonal prism: it is shown in Fig. 5.11, which gives also the conventional notation for symmetry points. The free electron energy curves, calculated in the same way as those for the f.c.c. lattice (Fig. 5.2), are illustrated in Fig. 5.12. The bands are classi-

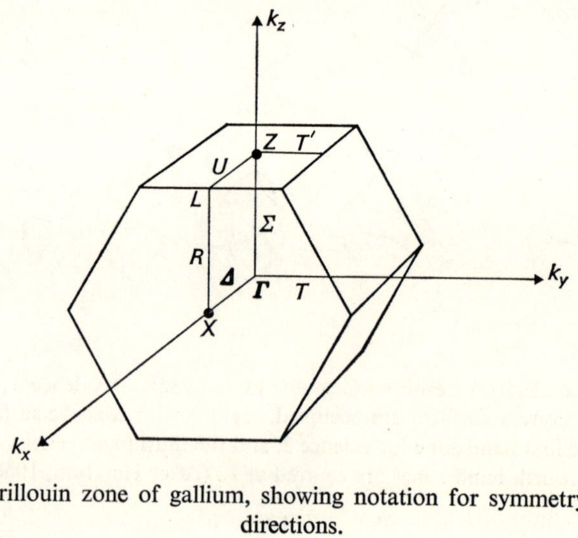


Fig. 5.11. Brillouin zone of gallium, showing notation for symmetry points and directions.

fied according to group theory (Section 6.1). The Fermi level is shown and it is apparent how widely varying (from 2 to 9) is the number of bands occupied in different parts of the Brillouin zone. For example, as we go from $\Gamma$ to the zone boundary along the T-direction, this number is first nine, then seven, then

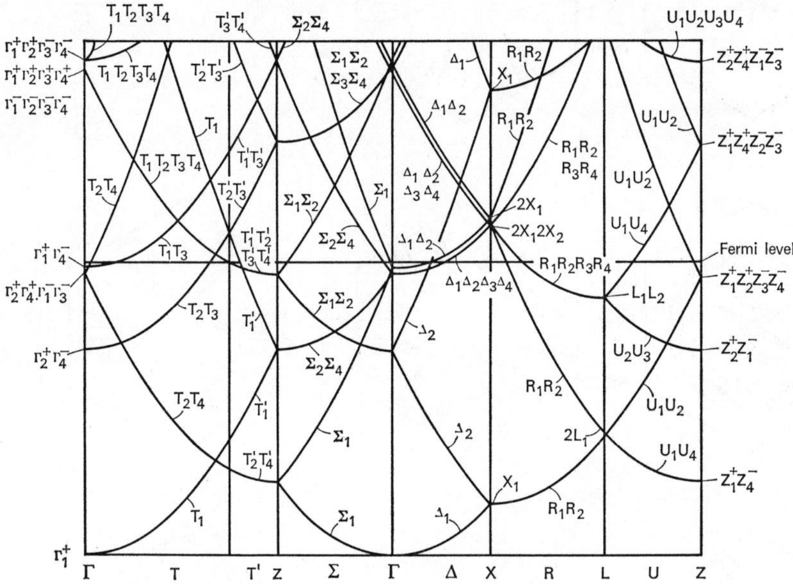


Fig. 5.12. Free electron energy bands in gallium.

five, then three. Correspondingly the Fermi surface is enormously complicated because there is a 2–3 surface, separating the regions where the number of occupied bands is respectively 2 and 3, a 3–4 surface ... a 8–9 surface. Each of these surfaces, which in general are independent of each other, can be concerned in the de Haas – van Alphen effect, magnetoresistance, etc. The sections of these Fermi surfaces with planes perpendicular to the crystal axis are illustrated in Figs. 5.13 and 5.14 (Slater, Koster and Wood, 1962). Of course not all the details can be correct, but more accurate treatments have led to the same sort of surfaces.

## 5.3 Nearly free electrons

To improve on completely free electrons, we introduce a small periodic potential $V(\mathbf{r})$ that may be treated as a perturbation (nearly free electron – NFE – approximation). The most noticeable effect of restoring the periodic

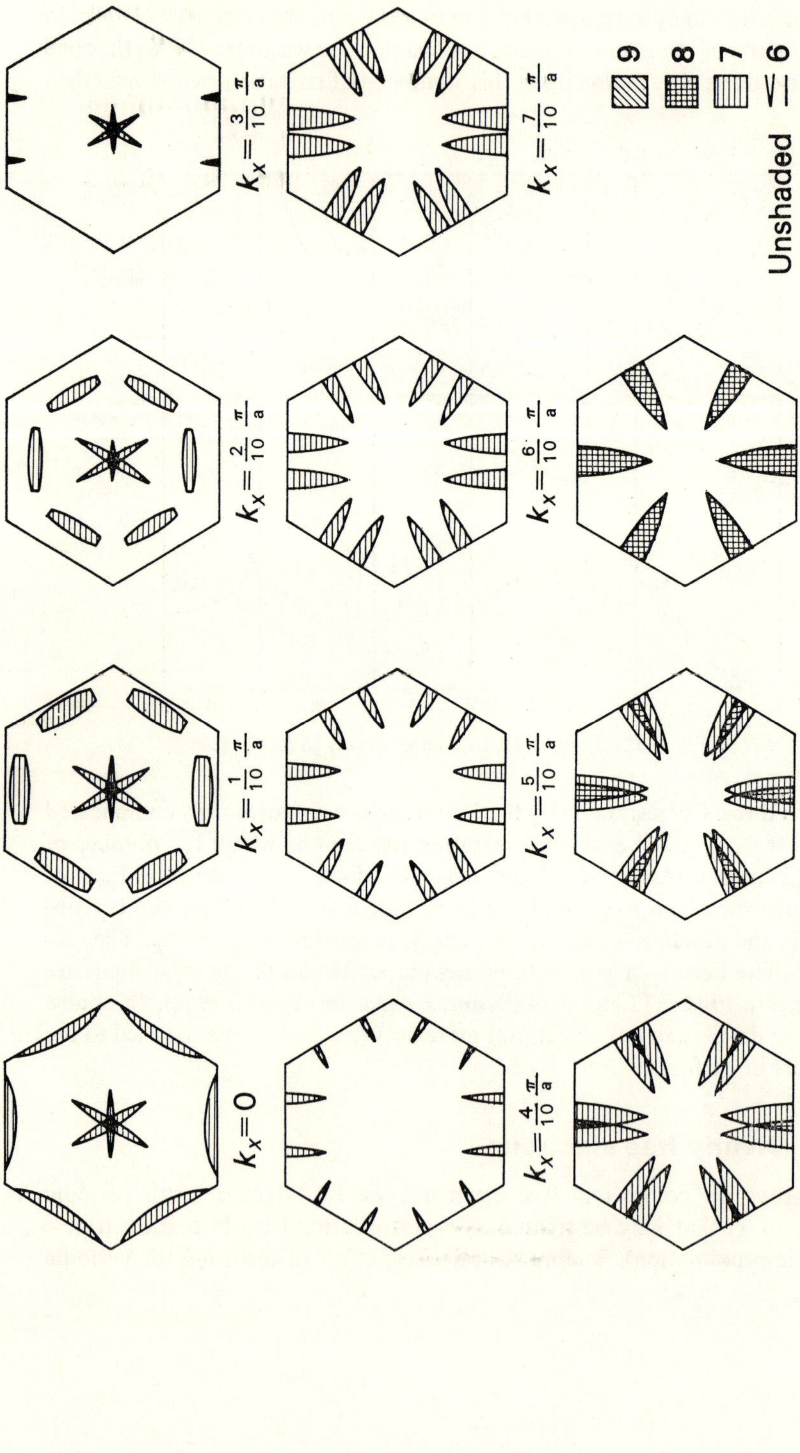


Fig. 5.13. Intersections of Fermi surfaces with planes from $k = 0$ to face of Brillouin zone. 6–7, 7–8 and 8–9 surfaces. (After Slater, Koster and Wood, 1962)

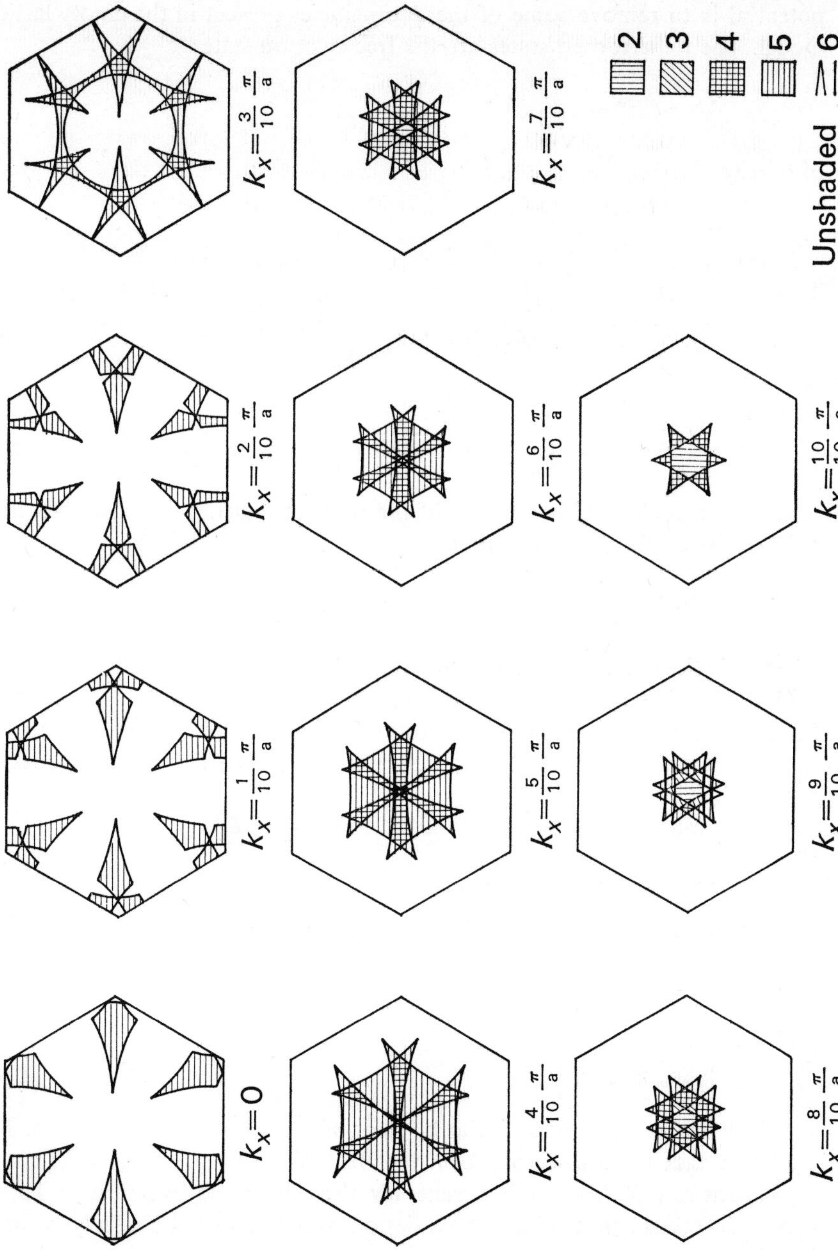


Fig. 5.14. As Fig. 5.13. 2–3, 3–4, 4–5, and 5–6 surfaces. (After Slater, Koster and Wood, 1962)

potential is to remove some of the degeneracies present in the empty lattice model. The unperturbed states are the free electron states

$$|\mathbf{k}\rangle = \Psi^0_{\mathbf{k}} = \frac{1}{\sqrt{\Omega}} e^{i\mathbf{k}.\mathbf{r}}$$

of energy

$$E^0_{\mathbf{k}} = \frac{\hbar^2 k^2}{2m_0}$$

To second order

$$E(\mathbf{k}) = E^0(\mathbf{k}) + \langle \mathbf{k}|V|\mathbf{k}\rangle + \Sigma_{\mathbf{k}'} \frac{|\langle \mathbf{k}|V|\mathbf{k}'\rangle|^2}{E^0_{\mathbf{k}} - E^0_{\mathbf{k}'}} \tag{5.8}$$

But

$$\langle \mathbf{k}|V|\mathbf{k}'\rangle = \frac{1}{\Omega} \int_{\text{unit cell}} e^{[i(\mathbf{k}'-\mathbf{k}).\mathbf{r}]}\, V(\mathbf{r})\, d\mathbf{r}$$

vanishes unless $\mathbf{k} - \mathbf{k}'$ is a vector of the reciprocal lattice, $\mathbf{K}_i$, and then it is the Fourier coefficient $V(\mathbf{K}_i)$, thus

$$E(\mathbf{k}) = E^0(\mathbf{k}) + \bar{V} + \Sigma_{\mathbf{K}_i \neq 0} \frac{|V(\mathbf{K}_i)|^2}{E^0_{\mathbf{k}} - E^0_{\mathbf{k}'}} \tag{5.9}$$

where $\bar{V}$ is the average value of the potential.

(5.9) is a good approximation if
(a) the coefficients $V(\mathbf{K}_i)$ tend to zero rapidly as $\mathbf{K}_i$ increases; we assume this for the moment and we shall see that 'pseudopotentials' can be introduced having this property;
(b) there must be no degeneracy of the form $E^0_{\mathbf{k}} = E^0_{\mathbf{k}'}$.
If

$$\mathbf{k}' = \mathbf{k} + \mathbf{K}_i$$

the condition

$$E^0_{\mathbf{K}} = E^0_{\mathbf{K}'}$$

means geometrically

$$|\mathbf{k}| = |\mathbf{k} - \mathbf{K}_i| \tag{5.10}$$

that is, $\mathbf{k}$ lies on the perpendicular bisector of the reciprocal lattice vector $\mathbf{K}_i$; we know that these bisectors form the Brillouin zone boundaries; (5.10) is equivalent to (2.42). There is degeneracy also where two free electron bands cross. (5.9) thus is not valid when $\mathbf{k}$ lies on or near a zone boundary, or near $\mathbf{k}$-points where bands cross. In these cases, the perturbation equations must be considered explicitly.

The method is then that of momentum eigenfunctions (Section 3.3) except that now only very few $V(\mathbf{K}_i)$ are considered. We illustrate it for a one-dimensional chain of period $a$ and for the potential

$$V(x) = 2V_0 \cos\frac{2\pi x}{a} \tag{5.11}$$

where $V_0$ is small compared to the kinetic energy of the electrons. The two lowest bands for positive $k$ have unperturbed wavefunctions and energies

$$\begin{aligned} \Psi_1^0 &= e^{ikx}, \qquad E_1^0 = k^2 \\ \Psi_2^0 &= e^{i\left(k-\frac{2\pi}{a}\right)x}, \qquad E_2^0 = \left(k - \frac{2\pi}{a}\right)^2 \end{aligned} \tag{5.12}$$

$E_1^0$ and $E_2^0$ become equal at $k = \dfrac{\pi}{a}$. In the presence of the perturbation, the two states are represented by linear combinations of the two waves

$$\Psi = A\, e^{ikx} + B\, e^{i\left(k-\frac{2\pi}{a}\right)x} \tag{5.13}$$

($A$ and $B$ to be determined) that diagonalise the energy. Substituting (5.13) into Schrödinger's equation

$$-\nabla^2\Psi + V\Psi = E\Psi,$$

we have

$$(k^2 - E)A\, e^{ikx} + \left[\left(k - \frac{2\pi}{a}\right)^2 - E\right] B\, e^{i\left(k-\frac{2\pi}{a}\right)x}$$
$$+ V_0\left(e^{\frac{2\pi i x}{a}} + e^{-\frac{2\pi i x}{a}}\right) \times \left(A\, e^{ikx} + B\, e^{i\left(k-\frac{2\pi}{a}\right)x}\right) = 0 \tag{5.14}$$

For (5.14) to hold, the coefficients of all the exponentials must vanish. The approximation consists in neglecting the terms

$$V_0 A\, e^{i\left(k+\frac{2\pi}{a}\right)x} \quad \text{and} \quad V_0 B\, e^{i\left(k-\frac{4\pi}{a}\right)x}$$

and this is justified since the bands represented by these wavefunctions are widely separated at $k = \pi/a$ from the two bands under consideration. The coefficients of the other exponentials give

$$\begin{aligned} (k^2 - E)A + V_0 B &= 0 \\ V_0 A + \left[\left(k - \frac{2\pi}{a}\right)^2 - E\right] B &= 0 \end{aligned} \tag{5.15}$$

These equations are in fact the equations (3.17) within the present approximation. They have non-zero solutions for the momentum eigenfunctions $A$, $B$ only if the determinant of the coefficients vanishes, or

$$E^2 - \left[k^2 + \left(k - \frac{2\pi}{a}\right)^2\right]E + k^2\left(k - \frac{2\pi}{a}\right)^2 - V_0^2 = 0 \qquad (5.16)$$

At the band edge $k = \frac{\pi}{a}$,

$$E = \frac{\pi^2}{a^2} \pm V_0 \qquad (5.17)$$

and the two symmetrised wavefunctions belonging to these two levels are the symmetric and antisymmetric combinations of $\Psi_1^0$ and $\Psi_2^0$. These two functions are real (standing waves due to the Bragg reflection). For the lower band

$$\Psi_s = C \cos \frac{\pi}{a} x$$

for the higher band

$$\Psi_a = C \sin \frac{\pi}{a} x$$

Thus the periodic potential separates the two bands, degenerate at the band edge in the free electron model, introducing an energy gap which, if the two free electron bands are characterised by wave numbers $K_m$, $K_n$ $\left(0, \frac{2\pi}{a}\right.$ in the $\left.\text{example}\right)$ is twice the Fourier coefficient $V(K_n - K_m)$. This is true also in three dimensions as can be seen, for instance, by assuming that the potential (5.11) acts in a cubic crystal. Schrödinger's equation is separable; the $y$- and $z$-dependent parts remain plane waves, the $x$-dependent equation is the same as in the one-dimensional example. Taking a f.c.c. lattice, with the Brillouin zone of Fig. 5.1, the energy gap across the hexagonal faces is, in the present approximation, $2V(\mathbf{K}_1)$, where

$$\mathbf{K}_1 = (\pm 1, \pm 1, \pm 1)\pi/2l \ (l = \text{side of elementary cube})$$

and is constant over the whole face; the energy gap across the square faces is $2V(\mathbf{K}_2)$, where

$$\mathbf{K}_2 = (\pm 2, 0, 0)\frac{\pi}{2l}$$

If there is more than one atom per unit cell, the discontinuity in energy at a zone boundary may vanish in the nearly free electron approximation, and then it will in fact usually be small. For an approximate treatment, consider a potential which is the sum of atomic potentials $U$

$$V(\mathbf{r}) = \Sigma_{i,j}\, U(\mathbf{r} - \mathbf{R}_i - \mathbf{d}_j)$$

where $\mathbf{d}_j$ is a basis vector giving the distance from the origin of the $j$th atom in the unit cell. The $n$th Fourier component of $V$ is

$$V(\mathbf{K}_n) = C \int_{\text{one cell}} e^{-i\mathbf{K}_n\cdot\mathbf{r}}\, \Sigma_{i,j}\, U(\mathbf{r} - \mathbf{R}_i - \mathbf{d}_j)\, d\mathbf{r}$$

$$= C \int \Sigma_{i,j}\, e^{-i\mathbf{K}_n\cdot(\mathbf{r}-\mathbf{d}_j)}\, e^{-i\mathbf{K}_n\cdot\mathbf{d}_j}\, U(\mathbf{r} - \mathbf{R}_i - \mathbf{d}_j)\, d\mathbf{r}$$

$$= \Sigma_j\, U_j(\mathbf{K}_n)\, e^{-i\mathbf{K}_n\cdot\mathbf{d}_j}$$

where $\mathrm{U}_j(\mathbf{K}_n)$ is the $n$th Fourier coefficient of the potential due to the atoms in position $j$. If all the atoms are identical,

$$V(\mathbf{K}_n) = U(\mathbf{K}_n)\, \Sigma_j\, e^{-i\mathbf{K}_n\cdot\mathbf{d}_j} = U(\mathbf{K}_n) S_n$$

The quantity $S_n$ is called the structure factor and is important in determining the intensity of X-ray reflection from the plane perpendicular to $\mathbf{K}_n$. In particular, if, for a certain $\mathbf{K}_n$, $S_n = 0$, there is no X-ray reflection from these planes, and in the present approximation $V(\mathbf{K}_n) = 0$ gives no energy gap at the corresponding zone boundary.

As already mentioned in Section 2.9, sometimes, particularly in dealing with metals and alloys, it is useful to consider only those boundaries where a discontinuity of energy exists in the nearly free electron model. In this way, zones are constructed (Jones zones) which are different from, and larger than the Brillouin zones.

Consider now the one-dimensional example in more detail. For a full discussion see Sommerfeld and Bethe (1933), p. 386. By differentiating (5.16) it is immediately verified that $dE/dk = 0$ at the band edge and that $d^2E/dk^2$ is negative near the top of the lower band. Columns 2 and 3 of Table 5.3.1 give the energies of the two bands near the edge for $a = \pi$ and $V_0 = 0{\cdot}1$. Columns 4 and 5 give the values of $A(k)$, $B(k)$, normalised, for the lower band. These values are interchanged for the higher band. The average value of momentum

$$\langle p \rangle = A^2 k + B^2(k - 2)$$

for the lower band is given in Column 6. It may be easily verified that the velocity $dE/dk$ equals $2\langle p\rangle$ (the electron mass in the present units is $\frac{1}{2}$). The last column gives $\langle p\rangle$ for the higher band.

**Table 5.3.1**

| $k$ | $E_1$ | $E_2$ | $A$ | $B$ | $p_1$ | $p_2$ |
|---|---|---|---|---|---|---|
| 0·7 | 0·4817 | 1·6983 | 0·9966 | −0·0825 | 0·686 | −1·286 |
| 0·8 | 0·6277 | 1·4523 | 0·9925 | −0·1222 | 0·770 | −1·170 |
| 0·84 | 0·6904 | 1·3608 | 0·9886 | −0·1508 | 0·795 | −1.115 |
| 0·88 | 0·7544 | 1·2744 | 0·9806 | −0·1960 | 0·803 | −1·043 |
| 0·92 | 0·8177 | 1·1951 | 0·9612 | −0·2757 | 0·768 | −0·928 |
| 0·96 | 0·8735 | 1·1297 | 0·9013 | −0·4332 | 0·585 | −0·665 |
| 0·98 | 0·8927 | 1·1081 | 0·8281 | −0·5606 | 0·351 | −0·391 |
| 0·99 | 0·8981 | 1·1021 | 0·7733 | −0·6340 | 0·186 | −0·206 |
| 1·0 | 0·9 | 1·1 | 0·7071 | −0·7071 | 0 | 0 |

The table shows that, as $k$ increases, in the lower band near the edge the momentum of the electron decreases (negative effective mass) because of the increased probability of the presence of the state with negative momentum. What happens is that, as $k$ increases, more and more momentum is transferred to the lattice because of the Bragg reflection, so that there is an overall decrease of the momentum of the electron. Below $k = 0{\cdot}7$, formula (5.9) is correct to four figures.

The effective mass at the edge can also be calculated. Introducing the atomic constants, equation (5.16) becomes

$$E^2 - \frac{\hbar^2 E}{2m_0}\left(2k^2 + \frac{4\pi^2}{a^2} - \frac{4\pi k}{a}\right) + \frac{\hbar^4}{4m_0^2}k^2\left(k - \frac{2\pi}{a}\right)^2 - V_0^2 = 0$$

Near the edge we write $E = \dfrac{\pi}{a} - \varepsilon$, where $\varepsilon$ is small; then

$$E^2 - \frac{\hbar^2\pi^2 E}{m_0 a^2} + \frac{\hbar^4}{4m_0^2}\left(\frac{\pi^4}{a^4} - \frac{4\pi^2\varepsilon^2}{a}\right) - V_0^2 = 0$$

and introducing the energy gap $E_G = 2V_0$

$$E = \frac{\pi^2\hbar^2}{2m_0 a^2} \pm V_0 \pm \frac{\pi^2\hbar^4\varepsilon^2}{m_0^2 a^2 E_G}$$

so that

$$\frac{1}{m^*} = \frac{1}{\hbar^2}\frac{d^2E}{dk^2}$$

$$= \frac{1}{\hbar^2}\frac{d^2E}{d\varepsilon^2} = \pm\frac{2\pi^2\hbar^2}{m_0^2a^2E_G} \tag{5.18}$$

The effective mass is negative in the lower band. The corresponding hole mass is equal to the electron mass in the upper band, a result (Section 4.5) which is a fair approximation only for small effective masses. Introducing the free electron energy at the edge,

$$E_0 = \frac{\hbar^2}{2m_0}\frac{\pi^2}{a^2}$$

we have

$$\frac{m^*}{m_0} = \frac{E_G}{4E_0} \tag{5.19}$$

That $m^*$ is proportional to $E_G$ when there is interaction with only one band, follows also from the more general treatment of Section 4.5.

For example, the structure of the conduction bands of germanium and gallium arsenide is similar near $k = 0$. Gallium arsenide, with the zincblende structure, is the compound made up of the two elements which are on either side of germanium in the Periodic Table. The ratio of the forbidden gaps is $E_G(\text{GaAs})/E_G(\text{Ge}) = 1{\cdot}68$. The ratio of the effective masses at $k = 0$ is $m^*(\text{GaAs})/m^*(\text{Ge}) = 0{\cdot}072/0{\cdot}04 = 1{\cdot}8$.

We apply the nearly free electron approximation also to illustrate the fact that two bands of the same symmetry cannot cross, but become hybridised when the two values of the energy are close. Referring to Fig. 5.2a consider the two free electron bands crossing at $k_y = 0{\cdot}5$, $E = 2{\cdot}25$. For one of them the wave function is

$$\Psi_1 = e^{i(k_y-2)y}$$

and it is of type $\Delta_1$ (Section 6.1). The other band is degenerate and the different states are represented by different linear combinations of the waves

$$e^{i[x+z+(k_y-1)y]}, \qquad e^{i[-x+z+(k_y-1)y]}$$

$$e^{i[x-z+(k_y-1)y]}, \qquad e^{i[-x-z+(k_y-1)y]}$$

One of these states is of type $\Delta_1$, and this band cannot cross the $\Psi_1$ band. On the other hand, the other bands can cross it. The $\Delta_1$ band is represented (Section 6.1) by the wave function

$$\Psi_2 = \tfrac{1}{2}\, e^{i(k_y-1)y}\left(e^{i(x+z)} + e^{i(-x+z)} + e^{i(x-z)} + e^{i(-x-z)}\right)$$

The wave vectors of $\Psi_1$ and $\Psi_2$ differ by

$$\mathbf{k}_2 - \mathbf{k}_1 = \pm\mathbf{i} \pm \mathbf{k} + \mathbf{j}$$

so that we can assume that in the nearly free electron approximation we need consider only the Fourier coefficient of the potential $V(\mathbf{K}_1)$, which we denote simply by $V$. The equations (3.17) for the momentum eigenfunctions reduce to $(k_y \equiv k)$

$$\begin{aligned} [(k-2)^2 - E]A + 4VB &= 0 \\ 4VA + 4[2 + (k-1)^2 - E] &= 0 \end{aligned} \qquad (5.20)$$

and the condition for a non-zero solution gives the equation for $E$

$$E^2 - (2k^2 - 6k + 7)E + k^4 - 6k^3 + 15k^2 - 20k + 12 - 4V^2 = 0 \qquad (5.21)$$

Take for instance $V = 0{\cdot}05$: the two bands given by (5.21) are plotted in Fig 5.15 with, for comparison, the free electron bands. The normalised values of $A$, $B$ are given in Table 5.3.2 which makes clear the process of hybridisation.

For the validity of the nearly free electron model, the other condition is that the series of Fourier components $V(\mathbf{K}_i)$ should converge rapidly. This does not seem likely, since the field near any ion is strong, requiring for the potential Fourier components of short wavelength. Also, $\Psi$ must behave like

**Table 5.3.2**

| *Lower band* | | | *Higher band* | |
|---|---|---|---|---|
| $k$ | $A$ | $B$ | $A$ | $B$ |
| 0 | 0·194 | −0·981 | 0·999 | 0·049 |
| 0·1 | 0·239 | −0·971 | 0·998 | 0·061 |
| 0·2 | 0·309 | −0·951 | 0·997 | 0·081 |
| 0·3 | 0·427 | −0·904 | 0·993 | 0·117 |
| 0·4 | 0·638 | −0·770 | 0·979 | 0·203 |
| 0·5 | 0·894 | −0·447 | 0·894 | 0·447 |
| 0·6 | 0·979 | −0·203 | 0·638 | 0·770 |
| 0·7 | 0·993 | −0·117 | 0·427 | 0·904 |
| 0·8 | 0·997 | −0·081 | 0·309 | 0·951 |
| 0·9 | 0·998 | −0·061 | 0·239 | 0·971 |
| 1·0 | 0·999 | −0·049 | 0·194 | 0·981 |

an atomic orbital inside the core of any ion, and thus must in general oscillate rapidly, requiring high Fourier components also in its expansion. Because of these objections, the nearly free electron method was neglected for a long

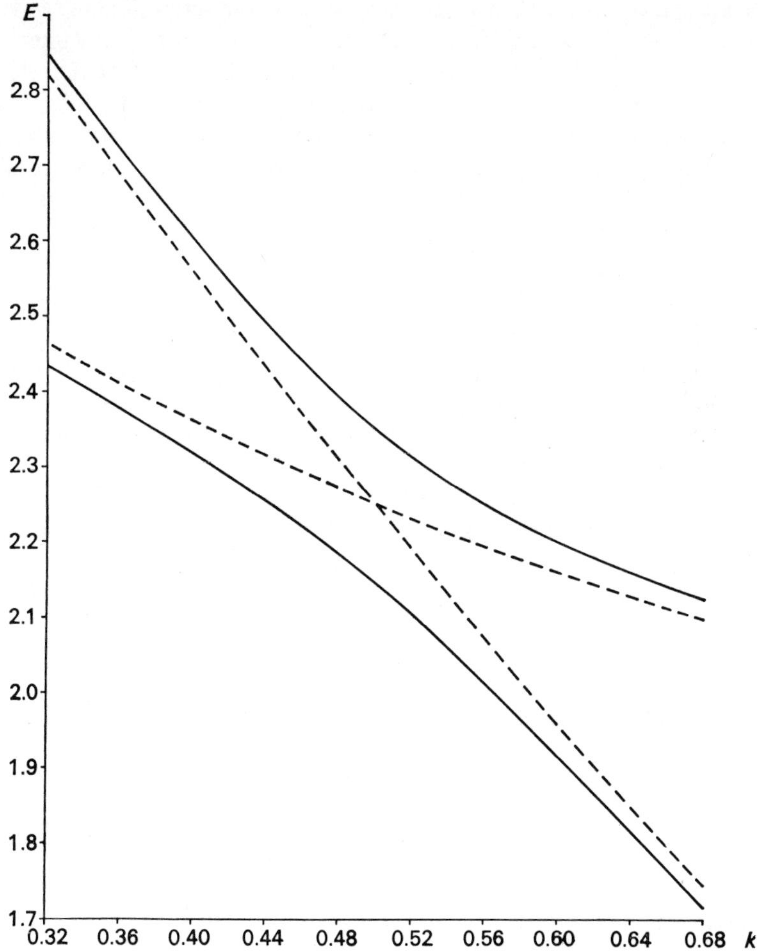


Fig. 5.15. Bands of the same symmetry do not cross. Dashed curves: bands in the free electron approximation.

time. However it became gradually apparent that it often gives better results than one would have expected. We shall see in the next Section why this is so and how the method can be made formally valid by the introduction of pseudo-potentials.

## 5.4 Orthogonalised plane waves and pseudopotentials

One of the main disadvantages of the methods for band structure calculation based on the use of plane waves is that the lowest eigenvalues correspond to core states, in which the electron is localised, so that a series of plane waves is a poor approximation to its wave function. Also, one is not usually interested in these states, but in states corresponding to higher roots of the secular equation for the energy. Following Herring (1940), it is then useful to make the plane waves, or the symmetrised plane waves (Section 6.2) orthogonal to the core states, so that the core energies are automatically excluded from the roots of the equation. The formalism of this method forms a good starting point for the introduction of pseudopotentials.

Let the strongly localised core wave functions be $b_t(\mathbf{r} - \mathbf{R}_i)$, where $t$ denotes the core state ($1s, 2s, 2p, \ldots$). Following Bloch, a function that satisfies the fundamental periodicity condition is

$$b_{t,\mathbf{k}}(\mathbf{r}) = \Sigma_{\mathbf{R}_i} e^{i\mathbf{k}\cdot\mathbf{R}_i} b_t(\mathbf{r} - \mathbf{R}_i) \tag{5.22}$$

(c.f. (3.23)). (5.22) describes the core state $t$, of energy $E_t$ in the crystal ($E_t$ may be different from the energy of state $t$ in the isolated atom). It may be considered as a very narrow band, fully occupied. We want now the higher states $\Psi_\mathbf{k}$ (valence band, conduction band – we omit the band index) to be orthogonal to $b_{t,\mathbf{k}}$ :

$$(b_{t,\mathbf{k}}, \Psi_\mathbf{k}) \equiv \int_{\text{unit cell}} b^*_{t,\mathbf{k}}(\mathbf{r})\Psi_\mathbf{k}(\mathbf{r})\, d\mathbf{r} = 0 \tag{5.23}$$

Since $\Psi_\mathbf{k}$ must behave like a free electron wave in the regions between the atoms, a plausible type of wave function for the higher states is

$$\Psi_\mathbf{k}(\mathbf{r}) \equiv X_\mathbf{k}(\mathbf{r}) = e^{i\mathbf{k}\cdot\mathbf{r}} - \Sigma_{t'} \beta_{t',\mathbf{k}} b_{t',\mathbf{k}}(\mathbf{r}) \tag{5.24}$$

where the coefficients $\beta$ must be determined so as to satisfy the orthogonality condition (5.23).

If the $b_{t,\mathbf{k}}(\mathbf{r})$ form an orthonormal system in the unit cell, (5.23), with $\Psi$ of the form (5.24), becomes

$$\int_{\text{unit cell}} b^*_{t,\mathbf{k}}(\mathbf{r})\, e^{i\mathbf{k}\cdot\mathbf{r}}\, d\mathbf{r} - \beta_{t,\mathbf{k}} = 0$$

so that we must take

$$\beta_{t,\mathbf{k}} = (b_{t,\mathbf{k}}, e^{i\mathbf{k}\cdot\mathbf{r}}) \tag{5.25}$$

This is the well known Schmidt procedure for the construction of sets of orthogonal functions. At **k**-points of symmetry, symmetrised combinations of plane waves must be used instead of a single plane wave. The function

$$X_{\mathbf{k}} = e^{i\mathbf{k}.\mathbf{r}} - \Sigma_t\,(b_{t,\mathbf{k}}, e^{i\mathbf{k}.\mathbf{r}})\,b_{t,\mathbf{k}}(\mathbf{r}) \qquad (5.26)$$

is called an orthogonalised plane wave (OPW). This function approaches a plane wave in the interatomic regions, where the $b_t(\mathbf{r})$ are small and has near the nuclei the rapidly varying character of atomic wave functions.

These orthogonalised plane waves are used as basic states for the wave function

$$\Psi_{\mathbf{k}}(\mathbf{r}) = \Sigma_{\mathbf{K}_i}\,a_{\mathbf{k}-\mathbf{K}_i}X_{\mathbf{k}-\mathbf{K}_i}(\mathbf{r}) \qquad (5.27)$$

where the coefficients $a_{\mathbf{k}-\mathbf{K}_i}$ are generalised momentum eigenfunctions; the $\mathbf{K}_i = 0$ term is identified with the predominant term in the expansion. The variational method is used to minimise the expectation value of the energy; remember however that the orthogonalised plane waves are not mutually orthogonal. The method is generally found to converge rapidly as more orthogonalised plane waves are added, and this is one of the indications that the nearly free electron method gives better results than we should expect.

To explain this, there is an argument first given by Phillips and Kleinman (1959). Suppose that $\Psi_{\mathbf{k}}$ is the exact wave function with some definite values of the $a_{\mathbf{k}-\mathbf{K}_i}$. Consider the function

$$\phi_{\mathbf{k}}(\mathbf{r}) = \Sigma_{\mathbf{K}_j}\,a_{\mathbf{k}-\mathbf{K}_j}\,e^{i(\mathbf{k}-\mathbf{K}_j).\mathbf{r}} \qquad (5.28)$$

in which simple plane waves replace the orthogonalised plane waves present in $\Psi_{\mathbf{k}}(\mathbf{r})$. Introducing (5.26) into (5.27) we have

$$\Psi_{\mathbf{k}}(\mathbf{r}) - \phi_{\mathbf{k}}(\mathbf{r}) = -\,\Sigma_t\,\Sigma_{\mathbf{K}_i}\,a_{\mathbf{k}-\mathbf{K}_i}b_{t,\mathbf{k}-\mathbf{K}_i}(\mathbf{r}) \int_{\text{unit cell}} b^*_{t,\mathbf{k}-\mathbf{K}_i}(\mathbf{r}')e^{i(\mathbf{k}-\mathbf{K}_i).\mathbf{r}'}\,d\mathbf{r}'$$

and, since

$$b_{t,\mathbf{k}-\mathbf{K}_i}(\mathbf{r}) = b_{t,\mathbf{k}}(\mathbf{r}) \text{ (see (2.39))},$$

$$\Psi_{\mathbf{k}}(\mathbf{r}) = \phi_{\mathbf{k}}(\mathbf{r}) - \Sigma_t(b_{t,\mathbf{k}}, \phi_{\mathbf{k}})b_{t,\mathbf{k}}(\mathbf{r}) \qquad (5.29)$$

This relation may be written

$$\Psi = (1 - P)\phi,$$

where $P$ is the projection operator that projects any state onto the core states, that is, that expresses it as a linear combination of core states.

Now substitute (5.29) into Schrödinger's equation $H\Psi = E\Psi$;

we have, omitting the suffix **k**,

$$H\phi - \Sigma_t (b_t, \phi)Hb_t = E\phi - E\,\Sigma_t (b_t, \phi)b_t$$

But

$$Hb_t = E_t b_t$$

thus

$$H\phi + \Sigma_t (E - E_t)(b_t, \phi)b_t = E\phi \tag{5.30}$$

We may look upon (5.30) as a new Schrödinger equation

$$(H + V_R)\phi = E\phi \tag{5.31}$$

where the operator $V_R$ means formally

$$V_R\phi = \Sigma_t (E - E_t)(b_t, \phi)b_t \tag{5.32}$$

Note that this formula gives the effect of $V_R$ only on $\phi$. It would not apply to other functions of position.

The function $\phi$ is smoother than $\Psi$, because it includes only plane waves, and not the rapid oscillations near the nucleus contained in the orthogonalised plane waves. It satisfies an equation whose Hamiltonian is

$$-\frac{\hbar^2}{2m_0}\nabla^2 + V + V_R$$

and which has the same non-core eigenvalues as the original equation. The effective potential

$$V_{\text{eff}} = V + V_R \tag{5.33}$$

contains the unknown function $\phi$. It is approximated by an ordinary potential (local pseudopotential) in most practical calculations. In principle it must be obtained through a self-consistent procedure.

The generalised momentum eigenfunctions $a_{\mathbf{k}-\mathbf{K}_j}$ satisfy a system of equations of the type of (3.16), except that the Fourier components of $V_{\text{eff}}$ appear instead of the Fourier components of $V$. $\phi$ is called a pseudowavefunction and $V_{\text{eff}}$ a pseudopotential. Although in practice $V_{\text{eff}}$ is treated as a local operator, it should be remembered that $V_R$ is in fact a non-localised operator (a non-linear integral operator):

$$V_R\phi = \Sigma_t (E - E_t)b_t(\mathbf{r}) \int b_t^*(\mathbf{r}')\phi(\mathbf{r}')\, d\mathbf{r}'$$

$$= \int \Sigma_t (E - E_t)b_t(\mathbf{r})b_t^*(\mathbf{r}')\phi(\mathbf{r}')\, d\mathbf{r}'$$

$$= \int V_R(\mathbf{r}, \mathbf{r}')\phi(\mathbf{r}')\, d\mathbf{r}'$$

with

$$V_R(\mathbf{r}, \mathbf{r}') = \Sigma_t (E - E_t) b_t(\mathbf{r}) b_t^*(\mathbf{r}')$$

Now the important point is that the Fourier components of $V_{\text{eff}}$ are small, except for the first few reciprocal lattice vectors, thus making valid the formalism of the nearly free electron model. The argument to show that $V_{\text{eff}}$ is small is that $V$ is negative, while $V_R$, containing $E - E_t$ and the square of an atomic orbital, is positive. There is some cancellation between them, and this cancellation can be improved by suitably changing the linear combination of core functions appearing in (5.32). Indeed the process of constructing $V_R$ is not unique. The non-core eigenvalues and eigenfunctions of the Hamiltonian $H + V_R$ are the same for any pseudopotential of the form

$$V_R\phi = \Sigma_t (F_t, \phi) b_t$$

where the $F_t$ are arbitrary functions. This follows from the fact that, as already remarked, the operator $V_R$ expresses $\phi$ as a linear combination of core states. But the non-core eigenfunctions are orthogonal to the sub-space of the core states, so that the addition of the operator $V_R$ to $H$, and any change in the coefficients of the linear combination of core states, make no difference to them. Thus by suitable choice of the $F_t$ the cancellation can be improved.

The non-uniqueness of the pseudopotential would not matter if we could obtain an exact solution of (5.30), because the same eigenvalues and true eigenfunctions would be obtained with any pseudopotential, even if the pseudowavefunctions would be different. However, perturbation theory is used and how good the results are, say to second order, depends on the choice of the pseudopotential. Several choices have been suggested; in particular, the smoothness of the pseudowavefunction has been suggested as a good criterion for the best pseudopotential. But it has been shown that this is so only at low energies and when a local pseudopotential is used (Harrison 1966, Pendry 1968). A possible choice, for example, is

$$F_t = -Vb_t$$

Then, let

$$V\phi = \sum_{t=1}^{n} a_t b_t \tag{5.34}$$

we have

$$a_t = \int b_t^* V\phi \, d\mathbf{r}$$

For an infinite sum (5.34) would be exact. For a finite sum over $n$ core states, there is an error, depending on the degree to which the $n$ core states form a complete set. Now, with $F_t = -VB_t$,

$$V_{\text{eff}}\phi = V\phi + \Sigma_t\left(\int -Vb_t^*\phi \, d\mathbf{r}\right)b_t$$

$$= V\phi - \sum_{t=1}^{n} a_t b_t$$

we have subtracted from $V\phi$ that part that can be expressed as a sum of core functions. If the bound states approximate well to a complete set in the region in which $V$ is large, the cancellation is good.

Often the pseudopotential is not calculated from first principles, but obtained or guessed from experimental measurements. Also pseudopotentials derived from experiment vary somewhat, depending on which data are used. But all the experimental pseudopotentials do not differ greatly, a fact which strongly supports the method.

The pseudopotential formalism based on orthogonalised plane waves is not the only possible one. The method can be formulated in terms of some other set of basis states. Indeed this type of argument is fruitful in many problems in the theory of metals and semiconductors and has made possible reliable calculations of quantities that before were intractable. In these problems the pseudopotential is better defined from the scattering properties of atoms or ions. When a plane wave of wave-vector $\mathbf{k}$ is scattered by an atom of radius $R$, the scattered wave can be expanded in Legendre polynomials; the radial dependence of the wave of angular momentum is, asymptotically, $\sin(kr + \eta_l)$, with $\eta_l = p_l\pi + \delta_l$, where $|\delta_l| < \pi/2$ and $p_l$ gives the number of radial nodes inside the sphere of radius $R$*. But the multiple of $\pi$ in the phase does not affect the scattering (see for instance Schiff 1949, p. 105), so that we can replace the true potential by a weaker one that will produce just the phase shift $\delta_l$, and all the scattering properties remain unchanged. This pseudopotential, generally different for different $l$, is suitable also for band calculations (cf. the method of scattered waves, Section 5.7.4). For our present purpose the introduction of pseudopotentials establishes the point that, because of orthogonality properties, the nearly free electron model is much better than one may think on first consideration. For a fuller discussion see Heine (1969, 1970); Cohen and Heine (1970).

## 5.5 Tightly bound electrons

When the electrons may be considered as fairly tightly bound to an atom, the nearly free electron model is not suitable; it is instead convenient to follow

* This is equal to the number of core states of angular momentum $l$.

Bloch's (1928) original suggestion of approximating the wave functions of electrons in crystals by linear combinations of atomic orbitals (LCAO), of the form

$$\Psi_{\mathbf{k},t}(\mathbf{r}) = N\,\Sigma_{\mathbf{R}_i}\, e^{i\mathbf{k}.\mathbf{R}_i}\, b_t(\mathbf{r}-\mathbf{R}_i) \tag{5.35}$$

where $N$ is a normalising factor, the term $e^{i\mathbf{k}.\mathbf{R}_i}$ ensures that the periodicity condition (2.9) is satisfied, and $b_t(\mathbf{r}-\mathbf{R}_i)$ is a normalised atomic wavefunction (1$s$, 2$s$, 2$p$ ...) centred at the lattice site $\mathbf{R}_i$; the method is an extension of the linear combination of atomic orbitals method for molecules, and is generally called the tight binding method. We have already used (5.35) to express the core functions in the orthogonalised plane wave method.

The band index is here a set of atomic quantum numbers. We know that, if the $b_t$ were the Wannier functions, the expansion (5.35) would be exact (Equation (3.23)). With atomic orbitals it is an approximation.

First we normalise $\Psi_{\mathbf{k},t}(\mathbf{r})$ in the unit cell. Omitting the suffix $t$, we have

$$N^2\,\Sigma_{\mathbf{R}_i,\mathbf{R}_j}\, e^{i\mathbf{k}.(\mathbf{R}_j-\mathbf{R}_i)} \int_{\text{unit cell}} b^*(\mathbf{r}-\mathbf{R}_i)b(\mathbf{r}-\mathbf{R}_j)\,d\mathbf{r} = 1$$

Writing

$$\mathbf{R}_i - \mathbf{R}_j = \mathbf{R}_m$$

this becomes

$$N^2\,\Sigma_{\mathbf{R}_i,\mathbf{R}_m}\, e^{-i\mathbf{k}.\mathbf{R}_m} \int_{\text{unit cell}} b^*(\mathbf{r}-\mathbf{R}_i)b(\mathbf{r}-\mathbf{R}_i+\mathbf{R}_m)\,d\mathbf{r} = 1$$

Now, integrating over unit cell and summing over $\mathbf{R}_i$ is the same as integrating over the whole crystal: then it is immaterial whether in the integrand we write $\mathbf{r}-\mathbf{R}_i$ or just $\mathbf{r}$, and the required normalisation condition is

$$N^2\,\Sigma_{\mathbf{R}_m}\, e^{-i\mathbf{k}.\mathbf{R}_m} \int_{\text{whole space}} b^*(\mathbf{r})b(\mathbf{r}+\mathbf{R}_m)\,d\mathbf{r} = 1 \tag{5.36}$$

The quantity

$$S_t(\mathbf{R}_m) = \int_{\text{whole space}} b_t^*(\mathbf{r})b_t(\mathbf{r}+\mathbf{R}_m)\,d\mathbf{r} \tag{5.37}$$

is called an overlap integral between atomic orbitals $t$: it is a measure of the overlapping of the atomic functions at atoms separated by $\mathbf{R}_m$. Thus the normalised wave function is

$$\Psi_{\mathbf{k},t}(\mathbf{r}) = \frac{\Sigma_{\mathbf{R}_i}\, e^{i\mathbf{k}.\mathbf{R}_i}\, b_t(\mathbf{r}-\mathbf{R}_i)}{[\Sigma_{\mathbf{R}_m}\, e^{-i\mathbf{k}.\mathbf{R}_m}\, S_t(\mathbf{R}_m)]^{\frac{1}{2}}} \tag{5.38}$$

Next we calculate the average value of the energy

$$\bar{E} = \int \Psi_{\mathbf{k}}^* H \Psi_{\mathbf{k}} \, d\mathbf{r}.$$

Indicating by $H(\mathbf{R}_m)$ the matrix element of $H$ between two atomic orbitals located on atoms separated by $\mathbf{R}_m$,

$$H_t(\mathbf{R}_m) = \int_{\text{whole space}} b_t^*(\mathbf{r}) H b_t(\mathbf{r} + \mathbf{R}_m) \, d\mathbf{r} \tag{5.39}$$

we find by the same methods used for the normalisation

$$\bar{E}_t(\mathbf{k}) = \frac{\Sigma_{\mathbf{R}_m} e^{-i\mathbf{k}.\mathbf{R}_m} H_t(\mathbf{R}_m)}{\Sigma_{\mathbf{R}_m} e^{-i\mathbf{k}.\mathbf{R}_m} S_t(\mathbf{R}_m)} \tag{5.40}$$

The integrals $H_t(\mathbf{R}_m)$ involved in (5.40) can be evaluated by writing

$$\begin{aligned} Hb_t(\mathbf{r} - \mathbf{R}_i) &= [-\nabla^2 + V(\mathbf{r})] b_t(\mathbf{r} - \mathbf{R}_i) \\ &= [-\nabla^2 + U(\mathbf{r} - \mathbf{R}_i)] b_t(\mathbf{r} - \mathbf{R}_i) \\ &\quad + [V(\mathbf{r}) - U(\mathbf{r} - \mathbf{R}_i)] b_t(\mathbf{r} - \mathbf{R}_i) \\ &= E_t b_t(\mathbf{r} - \mathbf{R}_i) + [V(\mathbf{r}) - U(\mathbf{r} - \mathbf{R}_i)] b_t(\mathbf{r} - \mathbf{R}_i) \end{aligned}$$

where $U(\mathbf{r} - \mathbf{R}_i)$ is the atomic potential centred on $\mathbf{R}_i$ and $E_t$ the atomic energy level belonging to the eigenfunction $b_t$. Thus

$$H_t(\mathbf{R}_i) = E_t S_t(\mathbf{R}_i) + \int_{\text{whole space}} b_t^*(\mathbf{r})[V(\mathbf{r}) - U(\mathbf{r} - \mathbf{R}_i)] b_t(\mathbf{r} - \mathbf{R}_i) \, d\mathbf{r} \tag{5.41}$$

The introduction of $U$, besides facilitating calculations, brings out the role played by the difference between the potential in the crystal and that in the isolated atom.

To discuss the main features of the method, assume for simplicity that all overlap integrals are negligible, except, of course, $S(0)$ which is unity if the $b_t$ are normalised. This simplification is reasonable when we consider tightly bound electrons (small orbits). With this approximation, the first term of (5.41) is $E_t$ for $\mathbf{R}_i = 0$ and vanishes for $\mathbf{R}_i \neq 0$. When $\mathbf{R}_i = 0$, the second term is the average value of $V - U$ calculated with the atomic wave functions. Since the $b_t$ decay rapidly away from the atom on which they are centred, it is legitimate to neglect all terms for which $\mathbf{R}_i$ is not the shortest lattice vector $\mathbf{R}_s$; that is, only the interaction with nearest neighbours is considered. Let then

$$g_t(\mathbf{R}_s) = \int_{\text{whole space}} b_t^*(\mathbf{r})[V(\mathbf{r}) - U(\mathbf{r})] b_t(\mathbf{r} - \mathbf{R}_s) \, d\mathbf{r} \tag{5.42}$$

(it is immaterial whether we consider $U(\mathbf{r} - \mathbf{R}_i)$ or $U(\mathbf{r})$): the average energy is

$$\bar{E}_t(\mathbf{k}) = E_t + \overline{(V - U)}_t + \Sigma_{R_s} e^{i\mathbf{k}.\mathbf{R}_s} g_t(\mathbf{R}_s) \tag{5.43}$$

the summation being over the nearest neighbours of the atom taken at the origin. As an example, suppose we are considering $s$-states, so that $b_t$ is spherically symmetric; then $g_t$ is the same for all equivalent $\mathbf{R}_s$ and

$$\bar{E}_0(\mathbf{k}) = E_0 + \overline{(V - U)}_0 + g_0 \, \Sigma_{R_s} e^{i\mathbf{k}.\mathbf{R}_s} \tag{5.44}$$

Note that in this case, in the region where the integrand in (5.42) is large the two $b$s have the same sign while $V - U$ is negative (Fig. 5.16) so that $g_0$ is a

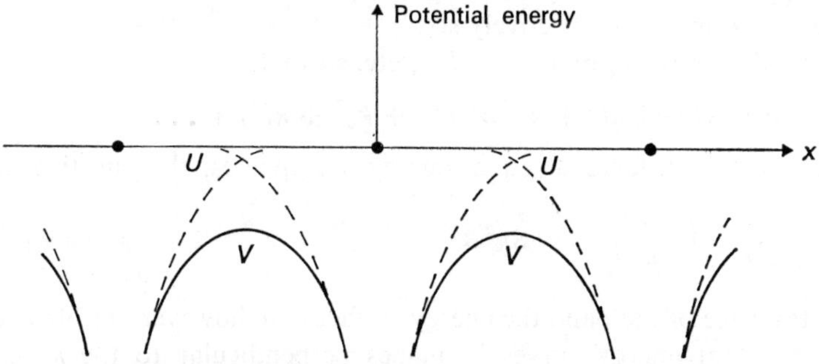


Fig. 5.16. Atomic potential, $U$, and crystal potential, $V$.

negative quantity, $g_0 = -B^2$. For a face-centred cubic lattice, if $a$ is the side of the elementary cube, the 12 shortest $\mathbf{R}_s$ are of type $(\pm\frac{1}{2}a, \pm\frac{1}{2}a, 0)$, so that, working out the sum in (5.44),

$$E = \text{constant} - 4B^2(\cos \tfrac{1}{2}k_x a \cos \tfrac{1}{2}k_y a + \cos \tfrac{1}{2}k_y a \cos \tfrac{1}{2}k_z a + \cos \tfrac{1}{2}k_z a \cos \tfrac{1}{2}k_x a)$$

For the example considered in treating the free electron model (Section 5.1), $a = 2\pi$, and thus

$$E(\mathbf{k}) = C - 4B^2(\cos \pi k_x \cos \pi k_y + \cos \pi k_y \cos \pi k_z + \cos \pi k_z \cos \pi k_x) \tag{5.45}$$

(5.45), which may be considered as a Fourier expansion for the energy, is then the tight binding $E(\mathbf{k})$ relation for an $s$-band in a face-centred cubic lattice. Similar simple formulae can be worked out in many cases. This theory brings out the fact that the width of a band is determined by the effective difference between the crystal potential $V$ and the atomic potential $U$.

In principle one can calculate the parameters, such as $B^2$, by taking suitable atomic orbitals, but in practice they are derived whenever possible from experimental data or from the results of other calculations. But even leaving them undetermined, a qualitative picture of the band is obtained. In the example we have, at prominent **k**-points:

$$E_\Gamma = C - 12B^2;\ E_L = C;\ E_K = C + 4B^2(2^{\frac{1}{2}} - \tfrac{1}{2});$$

along the diagonal of the square faces of the Brillouin zone, $E_X = E_W = \mathrm{C} + 4B^2$. Thus in this model the lowest level of an $s$-band in a f.c.c. monatomic lattice is at the centre of the Brillouin zone, while the highest level is indicated at some point on the diagonal of the square faces. If the maximum is at $X$, there are apparently six maxima, but effectively only three, because the centres of two opposite faces are equivalent points; if the maximum is at $W$, there are apparently 24 maxima, effectively six.

For small $k$, $\cos \pi k_x \simeq 1 - \frac{1}{2}\pi^2 k_x{}^2$, etc. so that

$$E = C - 12B^2 + 4\pi^2 B^2(k_x{}^2 + k_y{}^2 + k_z{}^2) + \ldots$$

near $k = 0$ the surfaces of equal energy are spheres, the effective mass is

$$\frac{1}{m^*} = \left(\frac{d^2E}{dk^2}\right)_{k=0} = 8\pi^2 B^2 \tag{5.46}$$

At the top edge of the band the energy surfaces are however anisotropic.

The constant energy curves in planes perpendicular to the $k_z$-axis are illustrated in Fig. 5.17. Writing $\pi k_x = x$, this refers to the particular case

$$E = -\cos x \cos y - \cos y \cos z - \cos z \cos x$$

The critical points are found by equating grad $E$ to zero; this gives the conditions

$$\begin{aligned} \sin x\,(\cos y + \cos z) &= 0 \\ \sin y\,(\cos z + \cos x) &= 0 \\ \sin z\,(\cos x + \cos y) &= 0 \end{aligned}$$

If

$$\sin x = \sin y = \sin z = 0$$

we have the point $\Gamma$, a minimum. If

$$\begin{aligned} \sin x = \sin y &= 0 \\ \cos x + \cos y &= 0 \end{aligned}$$

and equivalent relations, we have the diagonals of the square faces, an infinite number of maxima. The possibility

$$\begin{aligned} \sin x &= 0 \\ \cos z + \cos x &= 0 \\ \cos x + \cos y &= 0 \end{aligned}$$

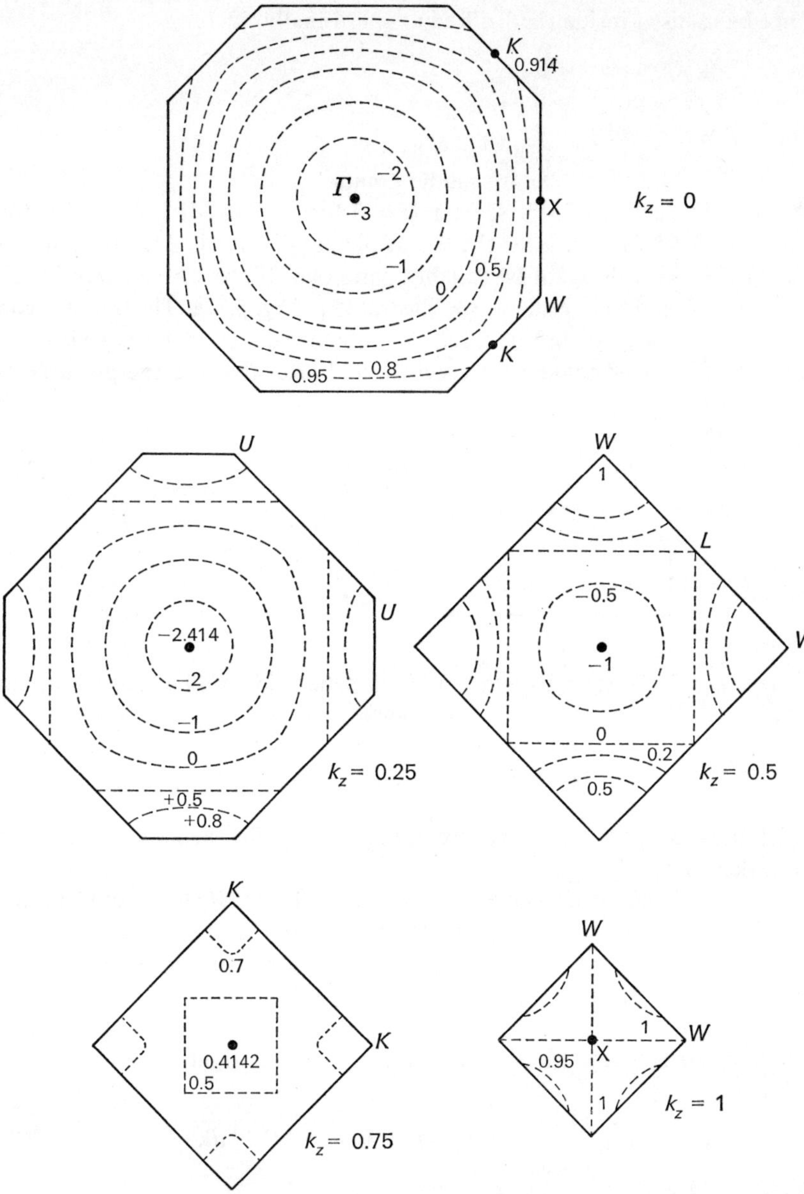


Fig. 5.17. Equienergy curves in planes perpendicular to the $k$ -axis for an s-band in the tight binding approximation. Face-centred cubic lattice.

$$E = -\cos x \cos y - \cos y \cos z - \cos z \cos x \qquad (x = \pi k \text{ etc.}).$$

cannot be realised inside the Brillouin zone. Finally, if

$$\cos y + \cos z = 0$$
$$\cos z + \cos x = 0$$
$$\cos x + \cos y = 0$$

we have the points $L$ which are saddle points.

As for the density of states, near $k = 0$ formula (2.54) holds, with the expression (5.46) for $m^*$. Near the top of the band, $1/m^*$ is a tensor, but still the $dn_s/dE$ curve will again be roughly parabolic. If the whole curve is calculated, it is found to have the form illustrated in Fig. 5.18a. The free electron density of states is given in Fig. 5.18b, for comparison. Here the curve is parabolic until the zone boundary is reached, and after that the energy surfaces

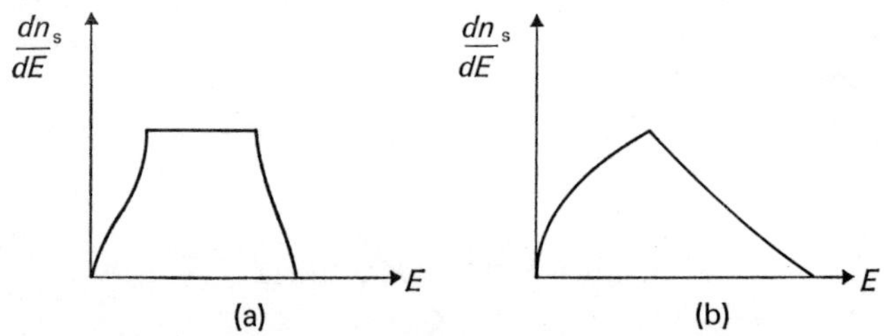


Fig. 5.18. Density of states (schematic) for the conduction band of a metal such as aluminium.
(a) Tight binding;
(b) Free electron approximation.

consist only of portions of spheres lying within the Brillouin zone, and $dn_s/dE$ decreases.

Note that if the denominator in (5.40) were 1, the $H_t(\mathbf{R}_m)$ would be the Fourier coefficients of the energy (3.22), which are the matrix elements of $H$ with respect to the Wannier functions. The denominator is equal to 1 for tightly bound electrons, showing again that in this limiting case the Wannier functions coincide with the atomic orbitals. In general, substituting (5.35) into (3.24) we have for the Wannier functions in the tight binding approximation

$$a(\mathbf{r} - \ {}_i) = \frac{1}{\Omega_B}\int_{BZ} N\,\Sigma_j\, b(\mathbf{r} - \mathbf{R}_j)\, e^{i\mathbf{k}\cdot(\mathbf{R}_j - \mathbf{R}_i)}\, d\mathbf{k}$$

and using (5.36), (5.37)

$$a(\mathbf{r} - \mathbf{R}_i) = \frac{1}{\Omega_B}\int_{BZ} \frac{\Sigma_j\, b(\mathbf{r} - \mathbf{R}_j)\, e^{i\mathbf{k}\cdot(\mathbf{R}_j - \mathbf{R}_i)}\, d\mathbf{k}}{(\Sigma_m\, S(\mathbf{R}_m)\, e^{-i\mathbf{k}\cdot\mathbf{R}_m})^{\frac{1}{2}}}$$

Now the tight binding method is useful when all overlap integrals are $\ll 1$, except $S(0)$, which, to first order is just 1; then

$$a(\mathbf{r}-\mathbf{R}_i) = \frac{1}{\Omega_B}\int_{BZ} [1 - \tfrac{1}{2}\Sigma_{m\neq 0} S(\mathbf{R}_m)\, e^{-i\mathbf{k}.\mathbf{R}_m}] \quad \Sigma_j b(\mathbf{r}-\mathbf{R}_j)\, e^{i\mathbf{k}.(\mathbf{R}_j-\mathbf{R}_i)}\, d\mathbf{k}$$

Because of the orthogonality of the plane waves in **k**-space, this reduces to

$$a(\mathbf{r}-\mathbf{R}_i) = b(\mathbf{r}-\mathbf{R}_i) - \tfrac{1}{2}\Sigma_{m\neq 0} S(\mathbf{R}_m)\, b(\mathbf{r}-\mathbf{R}_i-\mathbf{R}_m)$$

The integrals $H_t(\mathbf{R}_m)$ are generally difficult to evaluate because they are three centre integrals: $b(\mathbf{r})$ is centred about one atom, $b(\mathbf{r}+\mathbf{R}_m)$ is centred about another atom, and $H$ will involve potentials centred about third atoms.

Having evaluated the $E(\mathbf{k})$ relation with one type only of atomic orbitals, the next step is to consider wave functions made up of several $b_t$, representing different atomic orbitals, and use them in a variational procedure. These calculations are cumbersome and, for electrons not tightly bound, the convergence is poor. The method cannot be applied to excited states, since it is not possible to characterise them by localised states of electronic excitation. Each such state is in resonance with excited states localised at equivalent points in the lattice. The excited state, of crystal momentum **k**, must be a linear combination of these localised excitations. Frenkel (see, for instance, Knox (1963)) called these running states of excitation 'excitons'; we have here another quasi-particle and wave packets can be constructed to represent it. Wannier suggested another model for the exciton: it can be considered as a conduction band electron and a valence band hole bound together in a structure similar to positronium. The pair, whose separation may be quite large compared to interatomic distances, travels through the crystal in a state of total crystal momentum **k**. The Frenkel and the Wannier pictures may be considered as extreme cases; generally the picture is more complicated. Experimental information on excitons is gained mainly through a study of the fine structure which they produce in the optical absorption edge.

The tight binding method is chiefly used to gain in a simple way a qualitative picture of a band structure, much as the free electron method is used. One uses simply equation (5.40) for non-degenerate levels and considers as few orbitals as possible for degenerate cases. The method is also used for interpolating results obtained by more accurate methods at prominent **k**-points: the parameters appearing in the tight binding formulae are given values to fit the energy at these points, and the formulae are taken to hold over the whole of **k**-space.

Tight binding wave functions and orthogonalised plane waves may be combined in the interpolation procedure. This is useful for instance with transition metals where the $d$-bands, to be treated by the tight binding approximation, overlap the conduction band, to be treated by the nearly free electron approximation. The use of these procedures is equivalent to constructing a 'model Hamiltonian'. Again the parameters appearing in the model Hamiltonian are determined either from experimental data, or from results of accurate calculations at **k**-points of high symmetry. The results of these interpolations will be no more accurate than the underlying first principle calculation.

Several modifications of the tight binding procedure have been suggested which eliminate some of the disadvantages of the method. Ideally, if localised orbitals (site orbitals) are to be used, one would like to work with the Wannier functions, from which the exact $\Psi$s can be obtained through (3.23). However we have seen that the equations (3.34) for the Wannier functions are difficult to solve in practice. Löwdin (1950) has suggested linear combinations of atomic orbitals (orthogonalised atomic orbitals) which make orthogonal the functions centred on different atoms, thus eliminating overlap integrals; however the method is complicated, and Löwdin's functions are not solutions of the Hartree-Fock equations for the valence electrons.

For valence electrons in diamond-type crystals it is sometimes convenient to use orbitals localised at the covalent bonds (equivalent orbitals, very useful in the theory of molecules). It may be proved that these are the most localised orbitals that can be constructed.

## 5.6 Experimental information on band structures

Ideally, one would like to calculate electronic band structures of metals and semiconductors from first principles, using an absolute minimum of experimental information. Unfortunately, many of the quantities of interest, such as energy gaps of semiconductors, effective masses, detailed shape of the Fermi surface of metals, are sensitive to variations in the wavefunctions used, whatever these are, so that the ideal aim has never been completely achieved, and only a combination of theory and experiment can lead to the elucidation of all the details of a band structure. For a completely theoretical band structure to be accurate, total self-consistency must be achieved, electron-electron and electron-phonon interactions must be included, relativistic effects must be considered. The difficulty of the problem is illustrated by the many precise and careful calculations that have had little quantitative success.

On the other hand, with an empirical band model, there is always the danger that it is fitted to an incorrect or incomplete interpretation of experimental results, besides the possibility of experimental errors. However, between the two extremes of purely first principles calculations and purely empirical band models there is ample room for intermediate treatments (see, for instance, Herman et al., 1967), and these empirically adjusted first principles calculations can be very successful.

There is a great variety of experiments that provide information on band structures and related problems. For the determination of the Fermi surface of metals, we have seen (Section 4.6) that many details can be obtained from the de Haas-van Alphen effect, and by the oscillatory behaviour of other quantities, such as the magnetoresistance, with $B^{-1}$ at low temperature. The dependence on $B$ of the resistance indicates also whether a Fermi surface is, or not, multiply connected.

The surface resistivity of single crystals at high frequency (anomalous skin effect) provides information on the curvature of the Fermi surface, while with polycrystalline specimens it provides the total area of the surface. Information on the density of states through the conduction band is provided by soft X-ray spectra, though the interpretation of these spectra is not straightforward, and by the spectrum of photo-emitted electrons. The density of states at the Fermi level is found from the electronic contribution to the specific heat of the metal and from the paramagnetic susceptibility of the metal.

Cyclotron resonance in metals can be detected when $\mathbf{B}$ is parallel to the surface of the crystal. The electrons near the surface describe helical orbits with the axis parallel to the surface; they enter the skin depth, where the variable electric field is confined, once in each revolution. If the field accelerates them each time, they gain energy from the field and a resonance is observed (Azbel'-Kaner resonance, see Lifshitz, Azbel' and Slutskin, 1961). There is a resonance every time that $\omega$ satisfies

$$\omega = n\omega_c = n\frac{eB}{cm^*}$$

$m^*$ can thus be determined. A complication, which however is not very troublesome, is that, since $m^*$ is different in different orbits, the resonance frequency depends on the orbit.

Also magneto-acoustic phenomena, mainly the oscillations in the attenuation of ultrasonic waves when an applied magnetic field varies, provide information on the geometry of the Fermi surface. The absorption is a periodic function of $1/\mathbf{B}$ and the period gives, in principle, the diameter of the Fermi surface between points in $\mathbf{k}$-space where the velocity of the electrons is parallel to the electric field set up by the sound wave.

For semiconductors, magnetoresistance provides information on the location and shape of band extrema. A magnetic field influences the resistance only when the drift velocity of some of the carriers is not parallel to $\mathbf{B}$. Thus the absence of longitudinal magnetoresistance in all directions indicates spherical energy surfaces. In silicon there are minima of the conduction band along the [1,0,0] and equivalent directions. Thus, if **B** is parallel to one of these directions there is no magnetoresistance, while an effect arises if **B** is, say, along the [1,1,1] direction. Results on the variation of magnetoresistance with direction are sometimes difficult to interpret because any anisotropy in the scattering of carriers also has an effect. Piezoresistance (fractional change of resistance per unit strain) also depends on the direction of the uniaxial stress with respect to the axes of the crystal. If there are several energy minima, the stress destroys the symmetry, and the 'valleys' have no longer the same energy: electrons may be transferred to valleys where they have a different mobility.

We have seen (Section 4.6) that cyclotron resonance is a powerful technique for the determination of effective masses. As with most resonance experiments, high accuracy can be achieved. However, cyclotron resonance can be observed only if $\omega_c\tau$ ($\tau$ = scattering time) is larger than 1. If $B = 10^4$ gauss, the mobility must be $\sim 10^4\ \mathrm{cm^2V^{-1}s^{-1}}$. Other magneto-optical effects do not suffer from this restriction. These consist in the determination of the variation with **B** of the optical absorption or of the reflectivity of the crystal, for all possible orientations of **B** with respect to the crystal axes and for all wavelengths and directions of polarisation. In these experiments the mass parameter for any orbit in the magnetic field is (4.67)

$$m_c^* = \frac{\hbar^2}{2\pi}\frac{dA}{dE}$$

For the case of isotropic, though non parabolic, bands, we can write

$$\frac{dA}{dE} = \frac{dA}{dk}\frac{dk}{dE}$$

then, since

$$A = \pi k^2, \qquad \frac{1}{m_c{}^*} = \frac{1}{\hbar^2 k}\left(\frac{dE}{dk}\right)_{E_F} \tag{5.48}$$

('optical' effective mass). (5.48) differs from the standard expression (4.33) for the effective mass, though the two expressions become equal for isotropic parabolic bands. There are analogous differences in the meaning of $m^*$ also for other band shapes. Note that the sample must be degenerate, so that one may speak of a Fermi energy $E_F$.

An average effective mass is obtained also by determining the frequency for which the reflectivity has a minimum in the absence of a magnetic field. This occurs when $\omega = \omega_p$, the plasma frequency:

$$\omega_p^2 = \frac{4\pi Ne^2}{\ast}$$

The density of carriers, $N$, is involved in these effects and must be determined by a separate experiment.

For measuring energy gaps, the oldest method is to use the variation with temperature of the electric resistance, or of the Hall constant $R$. As is well known, a plot of $RT^{\frac{3}{2}}$ versus $1/2kT$ should be a straight line of slope $E_G$. The gap obtained is that at 0 K, assuming that

$$E_G(T) = E_G(0) + CT$$

where $C$ is a constant. Since this formula must break down near $T = 0$, and the ideal formula for $R$ never holds exactly, there is always an uncertainty. Rough values of $m^*$ are also obtained from these experiments, but all transport effects are rather insensitive to variations in $m^*$.

The optical absorption edge gives $E_G$ directly, except that absorption edges are never completely sharp. Here one must distinguish the two cases of direct and indirect transitions. For direct transitions, the **k** vector of the electron excited from the valence to the conduction band remains constant except for the small change by the momentum of the absorbed photon. The wavenumber of a visible photon is $\sim 20\,000 \text{ cm}^{-1}$, while the size of the Brillouin zone is $\sim 10^8 \text{ cm}^{-1}$. For these transitions, the absorption coefficient near an edge has the form

$$\alpha = C(\hbar\omega - E_G)^{\frac{1}{2}} \tag{5.49}$$

so that, by measuring $\alpha$ for various $\omega$ near the edge, $E_G$ can be obtained. Indirect transitions occur when the edges of the valence band and of the conduction band are not at the same point of **k**-space. The absorption or emission of a phonon (quantised lattice vibration) is then required to conserve **k**. Phonon energies are $\ll \hbar\omega$ for visible light, so that practically all the transition energy is provided by the photon. For these transitions the absorption coefficient near the edge is the sum of two terms

$$\alpha_{\pm} = C(\hbar\omega - E_G \pm W_s)^2 \tag{5.50}$$

corresponding to the emission or absorption of a phonon of energy $W_s$. Since the exponents in (5.49) and (5.50) are different, it is possible to determine whether the transition is direct or indirect. Optical absorption can be used also to study transitions between different branches of a valence band, thus determining, for instance, spin-orbit splitting energies.

At frequencies below the edge, some absorption occurs because of the presence of excitons. Taking the Wannier model (Section 5.5), there are absorption lines which at low temperature are sharp. If the binding energy can be calculated, $E_G$ can be obtained accurately. In these experiments the absorption due to the free carriers must be subtracted. In the strong absorption region, transmission experiments are impossible. However, reflectivity experiments provide information, because peaks of reflectivity are found, corresponding to peaks in the absorption due to electronic transitions between bands near symmetry points. The absoption is higher here because near symmetry points $\text{grad}_K E$ is generally small and so the density of states is large. A powerful technique is to carry out absorption and reflectivity experiments in the presence of alternating stresses or electric fields. Well known a.c. techniques permit the plotting of differential absorption curves where one can locate precisely the energies at which rapid changes of absorption occur and at which there are maxima.

The variation of the absorption near the edge in a magnetic field also provides accurate values of $E_G$. So does the recombination radiation which is maximum at a wavelength corresponding to $E_G$, particularly if laser action has been achieved.

Also the study of the change of band structure with pressure reveals properties of the normal band structure. By these experiments one can also separate the two effects mentioned in Section 1.3 which cause a variation of $E_G$ with temperature.

## 5.7 Band structure calculations

In calculations from first principles, the problem divides into two parts: the setting up of the one-electron Hamiltonian, and the solution of the corresponding Schrödinger equation with the periodicity condition (2.9). The two parts are interrelated because the problem must be tackled from the self-consistent point of view. In recent years some fairly self-consistent potentials have been set up, as a consequence partly of the use of pseudopotentials, which has much reduced the number of Fourier components of the potential that must be considered, and partly of the coming into use of bigger electronic computors. It is clear that self-consistency can be achieved only by working out the wavefunctions, and hence the charge density, for a large number of **k**-points. In practice, often a plausible potential or pseudopotential is constructed without much concern over self-consistency. But even if a self-consistent calculation is carried out, one should start from a good initial potential.

### 5.7.1 Construction of the potential

Without considering the problem in detail, if, for instance, orthogonalised plane waves are used, there are five main terms:

1. The potential $-Ze^2/r$ arising from the net core charge (Z = 1 for Na, $Z = 2$ for Mg, etc.).

2. The core potential, that is the potential arising from the remainder of the nuclear charge (atomic number less $Z$), plus the potential arising from the core electrons. This core potential, coming from a neutral system, is localised in a small region and is obtained from Hartree-Fock calculations for atoms.
3. The exchange interaction between outer electrons and core electrons, generally expressed by Slater's exchange potential (1.4), often calculated by assuming free electrons.
4. The potential due to the outer electrons, that is the potential due to the charge density of single orthogonalised plane wave states.
5. A correction called screening which arises when several orthogonalised plane waves are used.

The sum of these five terms gives an 'atomic' potential $U(\mathbf{r} - \mathbf{R}_i)$ about each nucleus $\mathbf{R}_i$, and one takes

$$V(\mathbf{r}) = \Sigma_{\mathbf{R}_i} U(\mathbf{r} - \mathbf{R}_i) \tag{5.51}$$

The setting up of the potential is of course more complicated for polyatomic crystals. For a discussion of the construction of pseudopotentials for metals, see Harrison (1966).

The effect of exchange and correlation among conduction electrons has not been adequately treated even for a gas of free electrons. Pseudopotentials generally do not consider this effect explicitly. For a discussion of this problem, based on the Bohm-Pines theory of plasma oscillations, see Pines (1963). For free electrons this theory indicates that one must add to the kinetic energy a term $E_{\mathrm{BP}}(k)$ which depends on the magnitude of the wave vector $\mathbf{k}$ only. The addition of this term to the free electron bands, such as those of Fig. 5.2, considerably improves the free electron approximation for many metals.

The reasons why self-consistency is not always essential, is that it is found that a calculated band structure does not depend critically on the details of the potential. Most levels shift as a whole when parameters in $V(\mathbf{r})$ are altered. However, there are often a few levels, called sensitive levels, that move with respect to the others, and the choice of the potential becomes critical when some sensitive level is important. The shifting of one sensitive level causes, for instance, the difference between the structure of the conduction band of germanium and that of silicon, as is confirmed by a study of mixed Ge-Si crystals, where the shift can be followed as a function of relative composition.

The second part of the problem has attracted much attention, and a great variety of methods has been suggested (Pincherle, 1960; Alder 1968). Many have been shown to be of little practical value and only a few have remained in use. All methods, applied without approximations, should give the same results, but the point is that the complexity of the problem makes approximations inevitable, so that in each particular case the method which yields reliable results with the least labour should be used.

We have already reviewed some of these methods, those based on the two extreme points of view of nearly free electrons, suitable for simple metals, and that of tight binding, suitable for inner core electrons. For semiconductors and for most transition metals, more elaborate procedures are required. One of these is the method of orthogonalised plane waves (Section 5.4) and we shall now examine three more, the method of augmented plane waves, that of the '$S$-matrix' (or Korringa-Kohn-Rostoker (KKR), or scattered waves) and the cellular method. In most cases (orthogonalised plane waves, tight binding), the periodicity conditions are automatically satisfied, but Schrödinger's equation is satisfied only approximately. In the cellular method the opposite occurs.

The practical procedure, except for the cellular method, is that discussed in Section 2.4. The expansion (2.28) for $\Psi$ is used, in terms of 'basis functions' $\chi_i$ which may be orthogonalised plane waves, augmented plane waves, atomic orbitals, etc, and the variational procedure is employed to minimise the energy. In the equation (2.32) the energy may appear also implicitly, for example when the $\chi_i$ have been obtained by integration of Schrödinger's equation with a trial energy.

### 5.7.2 Cellular method

The crystal is divided into cells by drawing all planes bisecting and perpendicular to the lines joining each atom in the lattice to its neighbours. In monatomic lattices, these cells are the Wigner-Seitz cells. Each atom finds itself at the centre of a cell. In polyatomic crystals there are generally cells of different shape. Even in crystals like NaCl it may not be justified to take equal cubic cells, because the $Cl^-$ ion is larger than the $Na^+$ ion.

In each cell the potential is assumed to be spherically symmetric (deviations from spherical symmetry may be considered later as a perturbation). The wavefunction can then be separated in each cell in spherical polar coordinates, the angular part being expressed by means of associated spherical harmonics. $\Psi$ will be a sum of terms each containing one spherical harmonic multiplied by a radial function. It is possible to determine from group theory (Section 6.3) which spherical harmonics (or combinations of them) appear at each **k**-point; as with plane waves, at **k**-points of high symmetry fewer terms are present in the expansion. The radial functions are determined by integration of the radial equation with the assumed potential and trial values of the energy. They must have the appropriate number of zeros. One has then an expansion in each cell, with unknown coefficients. These are determined by 'matching' the solutions at the boundaries of adjacent polyhedral cells by making them satisfy appropriate boundary conditions. This is more complicated for polyatomic substances and when the cell has a large number of faces and a low symmetry. Taking a monatomic substance and two opposite faces of a cell, separated by a lattice vector **R**, the boundary conditions at the two equivalent points A, B (Fig. 5.19) are

$$\Psi(\mathrm{A}) = \Psi(\mathrm{B})\, e^{i\mathbf{k}\cdot\mathbf{R}}$$

$$\frac{\partial \Psi}{\partial n}(\mathrm{A}) = -\frac{\partial \Psi}{\partial n}(\mathrm{B})\, e^{i\mathbf{k}\cdot\mathbf{R}} \qquad (5.52)$$

These conditions can be re-stated to refer to values at a single point.

The boundary conditions are satisfied at a number of points: the number of equations must of course be equal to the number of unknown coefficients.

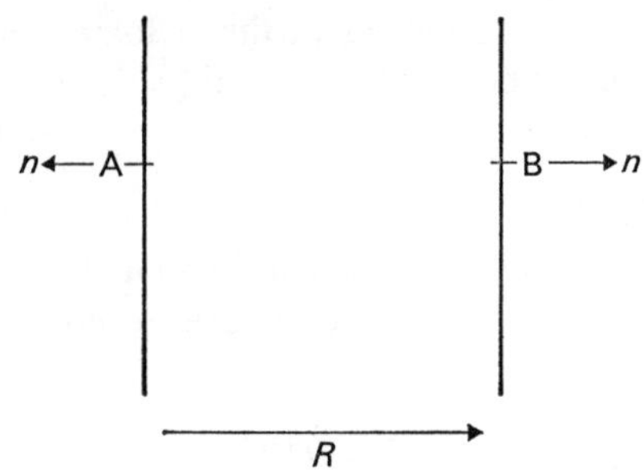


Fig. 5.19. Boundary conditions in the cellular method.

How many points can be considered, depends on the computational facilities. Least squares procedures, or variational procedures may alternatively be used, and they remove the arbitrariness inherent in point matching. In all cases, some compatibility relation is obtained, which is satisfied only for certain values of the trial energy, and it is easy to recognise which one of them is the required $E(\mathbf{k})$ for the band under examination.

The cellular method can yield reliable results only if a large number of terms is included in the expansion of $\Psi$. One must take care that the potential employed is not too inappropriate in the interatomic regions.

As a simple approximation, the polyhedral cell may be replaced by the sphere of equal volume. Then, because of periodicity and of inversion symmetry, the functions of even $l$ must have a vanishing normal derivative on the surface of the sphere, while those of odd $l$ must vanish on this surface. This approximation (Wigner and Seitz, 1933) works practically only for states near the bottom of an $s$-band. It can then be applied, for instance, to the determination of effective masses.

### 5.7.3 Augmented plane waves

The potential in a crystal in the regions between the ion cores is almost constant, a fact that is not generally taken into account in tight binding or cellular calculations. If it is assumed that the potential is perfectly constant in these regions, and is spherically symmetric within the atom cores ('muffin tin' potential), the problem of band structure calculation is much simplified. Consider a monatomic substance (the extension to polyatomic crystals is not difficult); let $R_s$ be the arbitrarily chosen radius of the spherical ion core: it may be the radius of the sphere inscribed in the Wigner-Seitz cell. In any case, the spheres are not supposed to be large enough to intersect. Schrödinger's equation can be solved exactly within each sphere as an expansion $\Psi_{\text{in}}$ in spherical harmonics, with a trial value $E_t$ of the energy and coefficients $a_l$. The radial functions are found, as in the cellular method, by numerical integration. In the region of constant potential (which can always be taken as zero) we have plane waves

$$\Psi_{\text{out}} = e^{i\mathbf{k}.\mathbf{r}}$$

These solutions, but not their gradients, are equated on the surface of each sphere by expanding $\Psi_{\text{out}}$ as a series of Legendre poynomials based on the atom under consideration

$$\Psi_{\text{out}} = \Sigma_l\, i^l(2l + 1)P_l\,(\cos\theta)\, j_l\,(kr) \tag{5.53}$$

where $j_l$ is a spherical Bessel function (for example $j_0(x) = \sin x/x$), and the angle $\theta$ is measured relative to $\mathbf{k}$. The two solutions could be made exactly equal by an infinite sum. In practice one keeps as many values of $l$ as allowed by the computational facilities. By equating $\Psi_{\text{in}}$ and $\Psi_{\text{out}}$, the coefficients $a_l$ are determined. There remains a discontinuity in grad $\Psi$. The combination of the two solutions is called an augmented plane wave (APW). With orthogonalised plane waves the two solutions are added; here we change from one expansion to the other on the dividing surface.

There are variations of this method. In one of them the expectation value of the energy, $\bar{E}$ is calculated for this augmented plane wave. There are contributions to $\bar{E}$ from the region inside the sphere, from the region outside, and also from the discontinuity in grad $\Psi$ at the surface. $\bar{E}$ is a function of $E_t$, and $E_t$ is varied to make $\bar{E}$ stationary: it is assumed that the stationary value is the best approximation to the correct energy. Now it is found that the value of $E_t$ that makes $\bar{E}$ stationary is equal to the stationary value $\bar{E}$ itself. Thus the part of the augmented plane wave within the atomic sphere is an eigenfunction of Schrödinger's equation belonging to an eigenvalue equal to the expectation value of $E$ for the whole crystal. The plane wave outside the sphere is not an exact solution, because $\bar{E}$ is not usually equal to $k^2$, but the error happens to be compensated exactly by the contribution to $\bar{E}$ due to the discontinuity in grad $\Psi$.

The relation between $E$ and $\mathbf{k}$ can be expressed analytically (see for instance Slater, 1965, or Callaway, 1964). If $V$ is the volume of the Wigner-Seitz cell minus the volume of the sphere of radius $R_s$, then

$$V(\bar{E} - k^2) = 4\pi \,\Sigma_l\, R_s^2(2l + 1)j_l^2(kR_s)\left[\frac{d}{dr} \log_e R_l(\bar{E}, r)\right]_{r=R_s} \qquad (5.54)$$

$\bar{E}$ is a multivalued function of $k^2$ because, as $E$ is changed, the logarithmic derivative of each radial function $R_l\,(\bar{E}, r)$ varies from $+\infty$ to $-\infty$ like a cotangent curve, being $\infty$ when a node of the radial function passes through $R_s$. Note that, as far as the $R_l(\bar{E}, r)$ are concerned, we require only their logarithmic derivatives at $r = R_s$. These derivatives can be obtained from the spectral terms of the free atom (quantum defect method, Ham, 1955).

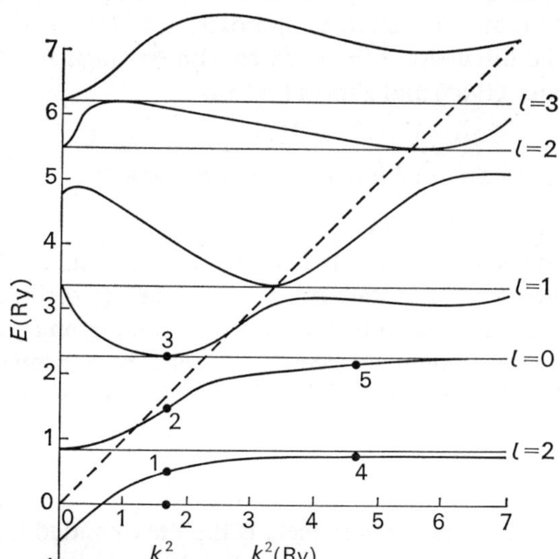


Fig. 5.20. Energy of augmented plane wave as a function of $k^2$, for copper (after Howarth 1955). The broken line indicates $E = k^2$.

The form of the curves one finds is illustrated in Fig. 5.20. The various curves correspond to solutions within the sphere with different predominant values of $l$ and different numbers of nodes. If the augmented plane waves were exact solutions, $\bar{E}$ would equal $k^2$, thus the solution to take as a first approximation is that which lies closer to the line $E = k^2$ at the **k**-point one considers. In Fig. 5.20, for $k^2 = k_0^2$, it is the solution corresponding to point 2. For a given $k^2$, the augmented plane waves of different energy are orthogonal, so that in this method orthogonalisation to the core functions is automatic. By plotting the same curves in the reduced zone scheme, an approximate band structure, spherically symmetric in **k**-space, is obtained. The next step is to take the augmented plane waves as basis functions for the variational procedure. One would combine, for any $\mathbf{k}_0$, functions with different $\bar{E}$ (points 1, 2, 3) and also functions whose plane wave part has vectors $\mathbf{k}_0 + \mathbf{K}_n$ (points 4, 5). At symmetric **k**-points, suitable linear combinations of augmented plane waves must be formed, and the spherical harmonics in the expansion within the sphere must also be combined so as to satisfy symmetry requirement (Section 6.3). The number of augmented plane waves is increased until the required energy values have converged satisfactorily. Often the convergence is fairly rapid. In the variational calculation, the restriction to a 'muffin tin' potential can be removed. (See Loucks, 1967.)

In Section 5.4 we discussed, on the basis of the orthogonalised plane wave method, why the nearly free electron approximation is much better than one would think. The discussion can be based also on augmented plane waves as shown by Slater (1965) and Ziman (1965).

### 5.7.4 Scattered waves and integral equation methods

The method of scattered waves is based on Ewald's (1916) discussion of the scattering of X-rays by crystals. It is essential in this method to assume a constant potential outside atomic spheres, and then (Korringa, 1947) every atomic sphere is considered to be the source of an outgoing wave $\phi$ of wave vector **k** and corresponding to electrons of energy $E$ propagating in the region of constant (zero) potential. Let $E = k_0^2$. The waves 'emitted' by two spheres $\mathbf{R}_i$ apart must satisfy the phase condition

$$\phi(\mathbf{r} + \mathbf{R}_i) = e^{i\mathbf{k}.\mathbf{R}_i}\,\phi(\mathbf{r}) \tag{5.55}$$

Now, at the surface of each sphere there is the wave emitted from the sphere itself, $\phi_{\text{out}}$, and the waves coming from all other spheres, combining in an incident wave $\phi_{\text{in}}$; for self-consistency $\phi_{\text{out}}$ and $\phi_{\text{in}}$ must be the same. But $\phi_{\text{in}}$ can be expressed in terms of $\phi_{\text{out}}$ by using (5.55). Expand them both in spherical harmonics, the coefficients being $a_{lm}$ for $\phi_{\text{out}}$, $b_{lm}$ for $\phi_{\text{in}}$. Equating $\phi_{\text{in}}$ and $\phi_{\text{out}}$ we get equations for $a_{lm}$, $b_{lm}$ where the energy enters as an unknown parameter. On the other hand $\phi_{\text{out}}$ and $\phi_{\text{in}}$ must be related as scattered and incident waves, that is through a scattering matrix depending on the potential $V(r)$ inside the sphere. If $V$ is spherically symmetric, scattering theory gives

$$a_{lm} = \tfrac{1}{2}b_{lm}[1 - e^{-2iq_l}]$$

$q_l$ being a phase shift which is a function of the energy and is determined by the condition that the solution $F_l(r)$ of

$$\frac{d^2F_l}{dr^2} + \left[k_0^2 - V(r) - \frac{l(l+1)}{r^2}\right]F_l = 0$$

outside the sphere should have the form

$$F_l(r) = C \sin(k_0 r - \tfrac{1}{2}\pi l + q_l)$$

If the potential is not spherically symmetric, each $a_{lm}$ is a linear combination of all the $b_{lm}$, the coefficients being the elements of the scattering matrix. In any case, one has two sets of equations for the $a_{lm}$, $b_{lm}$ which can be satisfied simultaneously only for certain values of the energy; **k** enters into the first set of equations and thus the dispersion relation is determined. The first part of the calculation is the determination of $\phi_{\text{in}}$ as the superposition of the waves coming from all atomic spheres and this requires the calculation of lattice sums (Ewald's sums) for chosen values of **k**, but does not depend on $V(\mathbf{r})$. Thus, apart from scaling factors, it is the same for all crystals with the same structure and can be carried out once and for all for any given structure. The second part, the scattering problem, is independent of **k**. In practice the calculation is carried out for a spherically symmetric potential and deviations from the 'muffin tin' form can be treated later by perturbation theory.

The integral equation approach looks quite different from the scattered waves method, but, if a variational procedure is adopted, and the potential is restricted to the 'muffin tin' type, as one must do in practice, the equations turn out to be identical.

Schrödinger's equation can be transformed into an integral equation by the general method of the Green function. For this reason the method is also called the Green's function method. Instead of

$$\nabla^2\Psi(\mathbf{r}) + E\Psi(\mathbf{r}) = V(\mathbf{r})\Psi(\mathbf{r}) \tag{5.56}$$

consider

$$\nabla^2 G(\mathbf{r}, \mathbf{r}') + EG(\mathbf{r}, \mathbf{r}') = -\delta(\mathbf{r} - \mathbf{r}') \tag{5.57}$$

where $G$ is the Green function satisfying the same periodicity condition as $\Psi$:

$$G_{\mathbf{k}}(\mathbf{r} + \mathbf{R}_i, \mathbf{r}') = e^{i\mathbf{k}.\mathbf{R}_i}\, G_{\mathbf{k}}(\mathbf{r}, \mathbf{r}') \tag{5.58}$$

The Green function may be expanded in terms of the eigenfunctions of the homogeneous equation

$$\nabla^2\Psi_n + E_n\Psi_n = 0$$

which form an orthonormal set:

$$G(\mathbf{r}, \mathbf{r}', E) = \Sigma_n A_n(\mathbf{r}')\Psi_n(\mathbf{r}) \tag{5.59}$$

Substituting (5.59) into (5.57) we obtain

$$\Sigma_n A_n(\mathbf{r}')\nabla^2\Psi_n(\mathbf{r}) + E\,\Sigma_n A_n(\mathbf{r}')\Psi_n(\mathbf{r}) = \delta(\mathbf{r} - \mathbf{r}')$$

or

$$\Sigma_n A_n(\mathbf{r}')(-E_n + E)\Psi_n(\mathbf{r}) = \delta(\mathbf{r} - \mathbf{r}')$$

Multiply by $\Psi_m^*(\mathbf{r})$ and integrate over the given domain (unit cell in our case); we obtain

$$A_m(\mathbf{r}')(E - E_m) = \Psi_m^*(\mathbf{r}')$$

and thus

$$G(\mathbf{r}, \mathbf{r}', E) = \Sigma_n \frac{\Psi_n^*(\mathbf{r}')\Psi_n(\mathbf{r})}{E - E_n}$$

Thus in the present case

$$G = \frac{1}{\Omega}\Sigma_{\mathbf{K}_n} \frac{e^{i(\mathbf{k}+\mathbf{K}_n).(\mathbf{r}-\mathbf{r}')}}{E - |\mathbf{k} + \mathbf{K}_n|^2} \tag{5.60}$$

It is immediately verified that

$$G^*(\mathbf{r}, \mathbf{r}') = G(\mathbf{r}', \mathbf{r})$$

and that (5.58) is satisfied. Multiplying now (5.56) by $G^*$ and the complex conjugate of (5.57) by $\Psi$, subtracting and integrating with respect to **r** over the unit cell, the

$$\int (G^*\nabla^2\Psi - \Psi\nabla^2 G^*)\, d\mathbf{r}$$

may be transformed into a surface integral which vanishes because of the periodicity conditions and one is left with (interchanging **r** and **r′** and omitting the suffix **k**)

$$\Psi(\mathbf{r}) = \int_{\text{unit cell}} G(\mathbf{r}, \mathbf{r}')V(\mathbf{r}')\Psi(\mathbf{r}')\, d\mathbf{r}' \tag{5.61}$$

which is the integral equation satisfied by $\Psi$. The energy appears in the Green function (5.60). The same equation can be written also in **p**-space using the momentum eigenfunctions $v(\mathbf{p})$ and the Fourier transform of the potential $V(\mathbf{p} - \mathbf{p}')$:

$$p^2 - E)v(\mathbf{p}) = \int V(\mathbf{p} - \mathbf{p}')v(\mathbf{p}')\, d\mathbf{p}'$$

Since for the periodic problem $V(\mathbf{p} - \mathbf{p}')$ is a sum of $\delta$-functions with coefficients $V(\mathbf{K}_n)$, we obtain again the equations (3.20).

Formally, (5.61) can be solved by successive approximations (iterations). A succession of iterations and variations should in principle yield rapidly the eigenvalues. However it is found that (5.61) is difficult to treat without restrictions on the potential. Kohn and Rostocker (1954) observed that the solution is much simpler with a 'muffin tin' potential because then the integration has to be carried out only within the atomic sphere. The procedure then is to replace the integral equation by a variational principle $\delta\Lambda = 0$, where,

$$\Lambda = \int_{\text{unit cell}} \Psi^*(\mathbf{r})V(\mathbf{r})\Psi(\mathbf{r})\, d\mathbf{r} - \iint_{\text{unit cell}} \Psi^*(\mathbf{r})V(\mathbf{r})G(\mathbf{r}, \mathbf{r}')V(\mathbf{r}')\Psi(\mathbf{r}')\, d\mathbf{r}\, d\mathbf{r}' \quad (5.62)$$

As usual, if $\Lambda$ is computed with some trial function $\Psi_t$, the error in the energy derived from $\delta\Lambda = 0$ is of second order compared to that of the trial function. Since all integrations are within the atomic sphere and $V(\mathbf{r})$ is assumed to be spherically symmetric, $\Psi_t$ can be expanded in spherical harmonics. Also $G$ is expanded in spherical harmonics and then the equations derived from $\delta\Lambda = 0$ are found to become identical to those of the scattered waves approach. The restriction to potentials that vanish outside the atomic sphere has also the advantage, as in the augmented plane wave method, that (5.62) can be converted to a form involving only integrals over the surface of this sphere and requiring only the logarithmic derivatives of the radial functions at $R_s$. A KKR pseudopotential can also be constructed.

Fully relativistic versions of the orthogonal plane wave, augmented plane wave and KKR methods have been developed, all however based on the free electron exchange approximation. Except for the lighter elements, relativistic corrections are important and a relativistic approach should be adopted for an accurate calculation.

# 6

# Classification of bands

## 6.1 Application of group theory

The classification of the bands is based on the space symmetry of the various states. The appropriate instrument is group theory, and a knowledge is assumed of the fundamental concepts of this theory (see for instance Heine, 1960; Cornwell 1969; Nussbaum, 1966), though we shall later discuss some of the ideas more relevant to band calculation. Group theory provides a sort of inventory of the possible forms of the bands at symmetry points in the Brillouin zone, specifically in terms of the permitted level degeneracies. Of course it cannot by itself predict the order of the levels.

We have already seen how the symmetry under translations leads to the classification of the electronic levels by means of **k**-vectors. In the language of group theory $\Psi_n(\mathbf{k}, \mathbf{r})$ satisfying

$$T_s \Psi_n(\mathbf{k}, \mathbf{r}) = e^{i\mathbf{k}.\mathbf{R}_s} \Psi_n(\mathbf{k}, \mathbf{r}) \tag{6.1}$$

is a basis function for an irreducible representation (see below) of the translation group. This is an Abelian group: all the operations commute and all the representations are one-dimensional. (6.1) diagonalises all the translation operators $T_s$ with the diagonal matrix element $e^{i\mathbf{k}.\mathbf{R}_s}$. Any given irreducible representation is characterised not by a specific **k**, but by the whole lattice of points $\mathbf{k} + \mathbf{K}_s$. In the reduced scheme, only the shortest of these vectors is considered.

Similarly, the classification of the bands at **k**-points with some symmetry is based on the symmetry properties at that point, more precisely on the group of operations (point group) that leaves the given point invariant. Thus the classification is generally different at different **k**-points. For instance, a band which is doubly degenerate at certain points, or along certain directions in **k**-space, may split into two non-degenerate bands at other points or along other directions. The bands generally split in going from a point of higher symmetry to a point of lower symmetry.

The symmetry operations of a crystal form a special kind of group called a space group. There are 230 different possible space groups. In many crystals, each element of the space group can be expressed as the product of an element of a translation group and an element of a point group which consists of either a rotation or of a rotation and an inversion. There are 32 possible point groups. For instance, the zincblende structure has a space group which consists of the f.c.c. translation group and the tetrahedral point group (see below). The space group of crystals with the diamond structure is not separable in this way, because of the presence of the screw type of symmetry (Fig. 2.12), symmetry which is absent in the zincblende structure because adjacent atoms are different.

One of the features of space groups is the existence of general and special positions within the unit cell. If an atom is located at a general position, there must be as many similar atoms in the unit cell as there are operations in the point group. But this is not so if the atom is at a special position, namely at a point or line of symmetry, because these points and lines are left unchanged by a number of operations in the group. For instance, if the atom is at the origin, there need be only one atom per unit cell.

The same happens for wave-vectors. If **k** terminates at a general position, there must be as many vectors of the same length $k$ as there are operations in the point group. All these wave-vectors are said to form a star. If **k** terminates at a special position, for example ($k_x$, 0, 0) in a cubic crystal, there are fewer vectors in the star (six in the example instead of 48). Two vectors on the boundary of the Brillouin zone which correspond to the same reduced wave vector are not considered as separate members of the star.

The operations of a point group do not generally commute. All the operations however commute with the one-electron Hamiltonian. In fact, if $R$ is any operation, it is immediately verified that

$$\nabla^2_{R\mathbf{r}} = \nabla^2_{\mathbf{r}}$$

and, by hypothesis,

$$V(R\mathbf{r}) = V(\mathbf{r}).$$

It follows that the energy is invariant under any operation of the group performed in **k**-space; the constant energy surfaces have the full crystal symmetry, with the practical consequence that it is necessary to determine the $E - \mathbf{k}$ relations only in a fraction of the Brillouin zone (1/48 for cubic structures). Also the momentum eigenfunctions, determined from the equations (3.16), have the full crystal symmetry.

Any operation of a space group induces rotations and reflections of the wave vector **k** in reciprocal space, as it does with the position vector in ordinary space. It is easy to see that the rotations and reflections produced in reciprocal space are the same as those produced in ordinary space by the inverse operation. Of course, there is nothing in reciprocal space corresponding to the translations.

For the purpose of classifying the bands at any **k**-point, one then considers point groups consisting of the operations that leave in the original position that particular **k**-point, or, for points on the boundary of the Brillouin zone, that move it to equivalent **k**-points. They are sub-groups of the point group of the crystal and must contain only operations that are symmetry operations of the crystal as a whole. For example, the truncated octahedron of Fig. 5.1 is the Brillouin zone of the zincblende structure and it would seem that the group to consider in the case of the point $\Gamma$ is the full cubic group. But 24 operations of the cubic group do not leave the zinc blende structure invariant: for these crystals the group to consider at $\Gamma$ is the tetrahedral group. At a general point in the Brillouin zone there is no symmetry whatever: the group of **k** consists of the single identity operation. At these points the bands can be classified only through the band index based on the energy.

Some general conclusions can be reached from a simple examination of point groups. For instance, if the group of the point $\mathbf{k}_0$ contains the inversion, the wavefunctions must have a definite parity (they must either remain unaltered or change sign under the inversion operation). Then $\Psi^*$ grad $\Psi$ is an odd function whatever the parity of $\Psi$. Thus $H'_{nn}$ in the expansion (4.43) of the energy in the neighbourhood of $\mathbf{k}_0$ must vanish. It must vanish, for example, at the points $\Gamma$ and $X$ in the Brillouin zone of a f.c.c. structure. On the other hand, if the point group does not contain the inversion, $H'_{nn}$ does not vanish. One thus expects to find maxima and minima of bands only at certain points of symmetry, or along symmetry axes. Instead of an extremum, the symmetry point may be a saddle point.

As an example of a point group, we list the 48 operations of the full cubic group:

(a) The identity $E$ leaving the cube in its original position.

(b) Rotations by $\frac{1}{2}\pi$ about an axis of fourfold symmetry, such as AA (Fig. 6.1), either to the left or to the right. There are six such rotations and they form the class $C_4$. Geometrically, two operations belong to the same class if it is possible to find a new coordinate system in which one operation is replaced by the other.

(c) Rotations by $\pi$ about the same axes. Three operations. Class $C_4^2$.

(d) Rotations by $\pi$ about an axis of two-fold symmetry, such as BB. Six operations. Class $C_2$.

(e) Rotations by $2\pi/3$, to the left or to the right, about an axis of threefold symmetry, such as DD. Eight operations. Class $C_3$.

(f) The inversion with respect to the origin, $J$.
(g), (h), (i), (j) Combinations of all the above rotations with the inversion (equivalent to reflections about symmetry planes). Twenty-three operations. Classes $JC_4$, $JC_4^2$, $JC_2$, $JC_3$.

The 48 elements of the cubic group are thus divided into 10 classes. If one denotes by $\pm x$, $\pm y$, $\pm z$ the faces of the cube, each operation can be specified

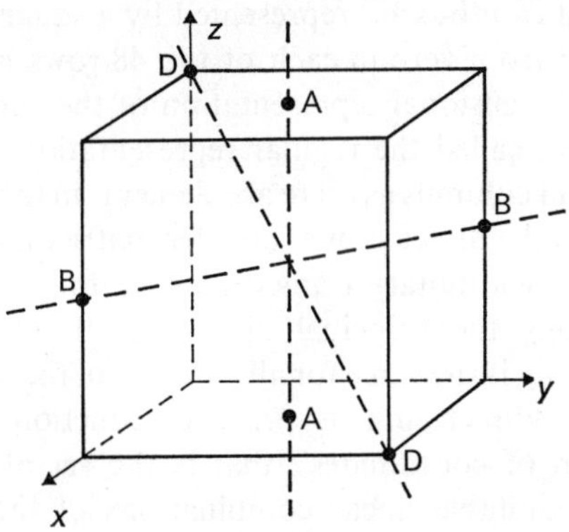


Fig. 6.1. Symmetry axes in a cube.

by indicating the faces replacing $+x$, $+y$, $+z$ after the operation has been performed. Thus the three $C_4^2$ operations are $(x, -y, -z)$, $(-x, y, -z)$, $(-x, -y, z)$.

If the elements of a group are replaced by square matrices which obey the same multiplication rules as the elements of the group, the matrices are called a representation of the group. The number of their rows and columns (the same for all matrices) is the dimension of the representation. Representations may have any number of dimensions. To construct a representation, the transformation properties of some linearly independent functions $\phi_1, \phi_2, \ldots, \phi_n$ are used. Let Q be one of the operations of the group. If

$$Q\phi_j = \Sigma_i D_{ij}(Q)\phi_i \tag{6.2}$$

and the $D_{ij}(Q)$ obey the same multiplication rules as the operations $Q$ of the group, the matrices $D_{ij}$ form a representation. For instance, the simplest case is when there is a function which is invariant, or only changes sign, under application of all the operations of the group $G$. All the matrices are then simply $(\pm 1)$; the representation is one-dimensional.

The functions $\phi_1, \ldots, \phi_n$ are said to form a basis for the representation. An alternative wording is that $\phi_1, \ldots, \phi_n$ transform according to the representation $D_{ij}$. To construct a representation of the full cubic group, consider the solutions of Schrödinger's equation at $\mathbf{k} = 0$ in the free electron model. If we consider the band labelled by the reciprocal lattice vector $(a, b, c)$, with $a \neq b \neq c$, there are 48 of these solutions with the same energy $a^2 + b^2 + c^2$. Starting for instance with the solution $e^{i(ax+by+cz)}$, by application of the operations of the cubic group, 48 different plane waves are generated, $e^{i(\pm ax \pm by \pm cz)}$, with all permutations of $x, y, z$, with either + or − sign. Call these waves $\phi_1, \phi_2, \ldots, \phi_{48}$. Any operation changes $\phi_i$ into one of the other 47, or into itself, and can thus be represented by a square matrix, with only one element different from zero in each of the 48 rows and columns. These matrices form a 48 dimensional representation of the cubic group.

This representation, called the regular representation, where the matrices have as many rows and columns as there are elements in the group, is reducible: it can be expressed as the direct sum of smaller matrices. That is, each matrix can be transformed, by a unitary transformation, to consist of a number of smaller matrices along the principal diagonal, all other matrix elements being zero. The splitting is the same for all matrices of the same representation. The symbol for this 'direct sum' is $\oplus$. The reduction can be effected by changing the 'system of coordinates', that is the set of basis functions. It means starting with suitable linear combinations of the plane waves. For instance the sum of the 48 waves with equal coefficients is clearly invariant with respect to any operation of the group. Each operation is then represented by the one-dimensional matrix (1). This one-dimensional irreducible representation exists for all groups.

If it is impossible to find any unitary transformation that will yield matrices of lower dimensions, the representation is said to be irreducible. In the case of the cubic group it is found that, by choosing suitable linear combinations of the original 48 waves, the $48 \times 48$ matrices can be expressed as the direct sum of four, generally different $3 \times 3$ matrices, each repeated three times, two, generally different $2 \times 2$ matrices, each repeated twice, and four $1 \times 1$ matrices. It is quite general that, when the regular representation is expressed as the direct sum of irreducible representations, an irreducible representation $n$-times degenerate is repeated $n$ times.

The number of irreducible representations of a group is equal to the number of classes in the group (ten in the example). The sum of the squares of the dimensions of the irreducible representations is equal to the number of elements of the group (in the example $4 \times 9 + 2 \times 4 + 4 \times 1 = 48$).

One of the most useful properties of representations is that the sum of the diagonal elements, or trace, of the matrices representing operations belonging to the same class is the same for all these operations, of course when the same representation is considered. Indeed, the matrices are not generally of immediate use, and most useful information is contained in their traces. This trace then characterises the class and is called the character of the representation for the given class. Any character refers thus to a certain class and a certain (reducible or irreducible) representation. When the irreducible representations have been found, characters can be displayed in a square Table, the rows being denoted by the irreducible representations and the columns by the classes. Since the identity $E$ is always represented by a unit matrix

$$\begin{pmatrix} 1 & 0 & 0 & \dots \\ 0 & 1 & 0 & \dots \\ 0 & 0 & 1 & \dots \\ . & . & . & \dots \end{pmatrix}$$

the character of the class $E$ must be equal to the dimension of the representation. Characters have the useful property that they are unaffected by unitary transformation of the matrices. In particular, the character of a reducible representation is, for each class, equal to the sum of the characters of the irreducible representations of which the reducible representation is the direct sum.

For the rules used to calculate characters, reference is made to standard textbooks.

Table 6.1.1 gives the characters for the $\Gamma$-group. The first column lists the

**Table 6.1.1**

*Character table of the $\Gamma$-group (full cubic group)*

| *Irreducible representations* | | *Basis functions* | *Classes* | | | | | | | | | |
|---|---|---|---|---|---|---|---|---|---|---|---|---|
| | | | $E$ | $C^2_4$ | $C_4$ | $C_2$ | $C_3$ | $J$ | $JC^2_4$ | $JC_4$ | $JC_2$ | $JC_3$ |
| $\Gamma_s$ | $\Gamma_1$ | $1$ | 1 | 1 | 1 | 1 | 1 | 1 | 1 | 1 | 1 | 1 |
| $\Gamma_p$ | $\Gamma_{15}$ | $x; y; z$ | 3 | −1 | 1 | −1 | 0 | −3 | 1 | −1 | 1 | 0 |
| $\Gamma_d$ | $\Gamma_{12}$ | $x^2 - y^2$; $z^2 - \frac{1}{2}(x^2 + y^2)$ | 2 | 2 | 0 | 0 | −1 | 2 | 2 | 0 | 0 | −1 |
| $\Gamma_{d'}$ | $\Gamma'_{25}$ | $yz; zx; xy$ | 3 | −1 | −1 | 1 | 0 | 3 | −1 | −1 | 1 | 0 |
| $\Gamma_f$ | $\Gamma'_2$ | $xyz$ | 1 | 1 | −1 | −1 | 1 | −1 | −1 | 1 | 1 | −1 |
| $\Gamma_{f'}$ | $\Gamma_{25}$ | $z(x^2 - y^2); x(y^2 - z^2)$; $y(z^2 - x^2)$ | 3 | −1 | −1 | 1 | 0 | −3 | 1 | 1 | −1 | 0 |
| $\Gamma_g$ | $\Gamma'_{15}$ | $xy(x^2 - y^2)$; etc | 3 | −1 | 1 | −1 | 0 | 3 | −1 | 1 | −1 | 0 |
| $\Gamma_h$ | $\Gamma'_{12}$ | $xyz(x^2 - y^2)$; $xyz[z^2 - \frac{1}{2}(x^2 + y^2)]$ | 2 | 2 | 0 | 0 | −1 | −2 | −2 | 0 | 0 | 1 |
| $\Gamma_i$ | $\Gamma_2$ | $x^4(y^2 - z^2)$ + etc | 1 | 1 | −1 | −1 | 1 | 1 | 1 | −1 | −1 | 1 |
| $\Gamma_l$ | $\Gamma'_1$ | $xyz[x^4(y^2 - z^2)$ + etc] | 1 | 1 | 1 | 1 | 1 | −1 | −1 | −1 | −1 | −1 |
| | | Number of elements | 1 | 3 | 6 | 6 | 8 | 1 | 3 | 6 | 6 | 8 |

irreducible representations, denoted with the suffixes $s, p, d, \ldots$ corresponding to the spherical harmonic of lowest $l$ appearing in the development, in terms of these functions, of the wavefunction associated with the given representation. The second column gives another widely used notation for the representations. The next column gives the simplest basis functions for the representation. The other columns give the characters for the various classes. Note the property, which is quite general, that, considering the numbers in each column of the Character Table as components of multi-dimensional vectors, any two of these vectors corresponding to different classes are orthogonal.

Table 6.1.2 gives the characters for the $\Delta$-group. The $\Delta$-direction is along an axis of fourfold symmetry in cubic crystals. The end points, $\Gamma$ and $X$, have

**Table 6.1.2**

*Character table of the $\Delta$-group*

| *Irreducible representations* | | *Basis functions* | *Classes* | | | | |
|---|---|---|---|---|---|---|---|
| | | | $E$ | $C_4{}^2$ | $C_4$ | $JC_4{}^2$ | $JC_2$ |
| $\Delta_s$ | $\Delta_1$ | $1$ | 1 | 1 | 1 | 1 | 1 |
| $\Delta_p$ | $\Delta_5$ | $x; y$ | 2 | −2 | 0 | 0 | 0 |
| $\Delta_d$ | $\Delta_2$ | $x^2 - y^2$ | 1 | 1 | −1 | 1 | −1 |
| $\Delta_{d'}$ | $\Delta'_2$ | $xy$ | 1 | 1 | −1 | −1 | 1 |
| $\Delta_g$ | $\Delta'_1$ | $xy(x^2 - y^2)$ | 1 | 1 | 1 | −1 | −1 |
| | | Number of elements | 1 | 1 | 2 | 2 | 2 |

however a higher symmetry. The operations that leave invariant a point along, say, the $z$-axis, are:

(a) The identity. Class $E$. One element.

(b) Rotation by $\pi$ about the $z$-axis. Class $C_4^2$. One element.

(c) Rotations of $\frac{1}{2}\pi$ about the $z$-axis in either direction. Class $C_4$. Two elements.

(d) Rotations by $\pi$ about the $x$ or $y$ axis followed by inversion. Class $JC_4^2$. Two elements.

(e) Rotation by $\pi$ about the axes of twofold symmetry in the $xy$-plane, followed by inversion. Class $JC_2$. Two elements.

The $\Delta$-group has thus eight elements forming five classes. Of the five irreducible representations, four are one-dimensional and one is two-dimensional.

A group in **k**-space which has the same Character Table as the $\Delta$-group is the $W$-group for the f.c.c. structure. Consider the $W_1$ point (Fig. 6.2) of coordinates $\pi/a(1, 2, 0)$, $a$ being the side of the cube of the face-centred structure. The group consists of operations that bring $W_1$ into an equivalent point, that is either into $W_2 \equiv \pi/a(1, -2, 0)$, displaced from $W_1$ by the reciprocal lattice vector $\mathbf{K}_2 = (0, -4\pi/a, 0)$, or into $W_{3,4} = \pi/a(-1, 0, \pm 2)$, displaced

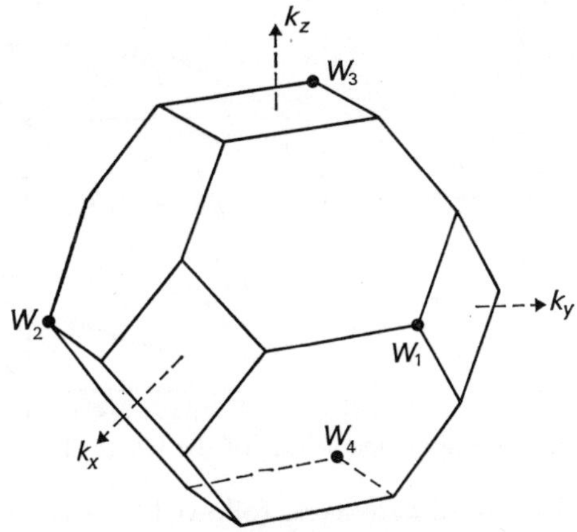


Fig. 6.2. Equivalent $W$-points.

from $W_1$ by the reciprocal lattice vectors $\mathbf{K}_1 = 2\pi/a(-1, -1, \pm 1)$. The group has eight elements divided into five classes, like the $\Delta$-group. The classes are

(a) Class $E$, the identity.

(b) Class $C_4^2$: rotation of $\pi$ about $k_x$. Interchanges $W_1$ and $W_2$, and $W_3$ and $W_4$. One element.

(c) Class $C_2$. Rotation of $\pi$ about one of the axes of twofold symmetry indicated in Fig. 6.1. Interchanges $W_1$ with either $W_3$ or $W_4$ and $W_2$ with either $W_4$ or $W_3$. Two elements.

(d) Class $JC_4^2$. Rotation of $\pi$ about $k_y$ or $k_z$ followed by inversion. Interchanges $W_1$ and $W_2$ leaving $W_3$, $W_4$ unchanged, or vice versa. Two elements.

(e) Class $JC_4$. Rotation of $\pm\frac{1}{2}\pi$ about $k_x$ followed by inversion. Interchanges $W_1$ with $W_4$ and $W_2$ with $W_3$, or $W_1$ with $W_3$ and $W_2$ with $W_4$. Two elements.

Although this group has the same Character Table as the $\Delta$-group, class $C_4$ is replaced by $JC_4$ and class $JC_2$ by $C_2$. Some of the basis functions are different: $\Delta_s$ and $\Delta_p$ are replaced by $W_s$ and $W^{(2)}{}_p$ with the same basis functions, but $\Delta_d$ is replaced by $W_f$ with basis function $xyz$, $\Delta_{d'}$ is replaced by $W_p^{(1)}$ with basis function $z$, and $\Delta_g$ is replaced by $W_d$ with basis function $xy$.

Another group important in semiconductors is the tetrahedral group, which consists of the 24 operations which send a regular tetrahedron into itself. Referring to Fig. 6.3, there are, besides the identity $E$, eight rotations by 120° about the axes of threefold symmetry (11, 22, 33, 44) (Class $C_3$ – corresponding to Class $C_3$ of the cubic group), and three rotations by 180° about axes of twofold symmetry ($x$, $y$, $z$ axes) (Class $C_2 - C_4^2$ of the cubic group); then six operations consisting of a rotation through 90°, to the right or to the

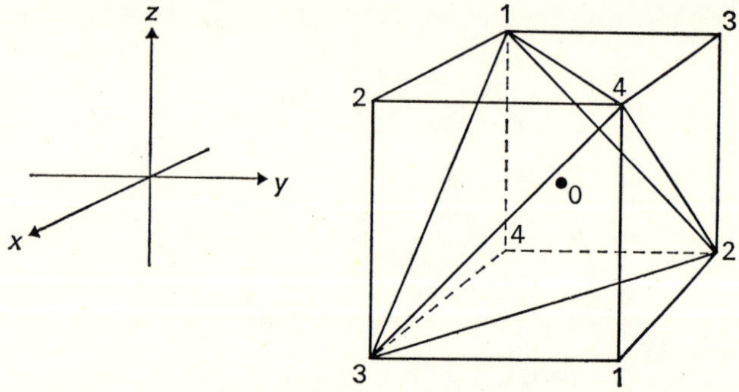


Fig. 6.3. Symmetry operations of the tetrahedral group.

left, about one of the coordinate axes, followed by a reflection in the plane perpendicular to this axis through the origin (Class $S_4 - JC_4$), and finally six reflections ($\sigma_d - JC_2$) in the six planes which pass through the centre and contain the six edges of the tetrahedron. This is the symmetry for crystals with the zincblende structure, for example the III-V semiconducting compounds such as GaAs. We know that for these crystals the Brillouin zone is the truncated octahedron pertaining to f.c.c. structures. Nevertheless, at $\mathbf{k} = 0$ the point group to consider is the tetrahedral group. Since there are five classes, there are five irreducible representations, two three-dimensional, of type $\Gamma_p$ and $\Gamma_{f'}$, one two-dimensional of type $\Gamma_d$, two one-dimensional, of type $\Gamma_s$ and $\Gamma_i$. The characters are given in Table 6.1.3.

For the diamond structure, the point group is the full cubic group. The difference with the simple f.c.c. structure is that, because of the presence of the screw axes, the diamond space group contains non-primitive translations in association with those operations of the group which are not also members of the tetrahedral point group; these operations have associated with them the translation $\mathbf{T} \equiv a(\frac{1}{4}, \frac{1}{4}, \frac{1}{4})$ which is the basis vector (Section 2.6) that produces the lattice. In other words, for these operations the members of the space group are the product of the operations of the cubic point group by the translation group $\mathbf{T} + \mathbf{R}_n$, while for the other operations they are the usual product of the point group by the translation group $\mathbf{R}_n$, where the $\mathbf{R}_n$ are the primitive translations of the f.c.c. Bravais lattice. The irreducible representations and characters are the same as for the cubic group (Table 6.1.1.)

**Table 6.1.3**

*Character table of the tetrahedral group*

| *Irreducible representations* | | *Basis functions* | *Classes* | | | | |
|---|---|---|---|---|---|---|---|
| | | | $E$ | $8C_3$ | $3C_2$ | $6\sigma_d$ | $6S_4$ |
| $\Gamma_s$ | $\Gamma_1$ | 1 | 1 | 1 | 1 | 1 | 1 |
| $\Gamma_i$ | $\Gamma_2$ | $x^4(y^2 - z^2) +$ etc | 1 | 1 | 1 | $-1$ | $-1$ |
| $\Gamma_d$ | $\Gamma_{12}$ | $x^2 - y^2; z^2 - \frac{1}{2}(x^2 + y^2)$ | 2 | $-1$ | 2 | 0 | 0 |
| $\Gamma_p$ | $\Gamma_{15}$ | $x; y; z$ | 3 | 0 | $-1$ | 1 | $-1$ |
| $\Gamma_{f'}$ | $\Gamma_{25}$ | $z(x^2 - y^2)$; etc | 3 | 0 | $-1$ | $-1$ | 1 |

We recall the fundamental theorems for the application of group theory to quantum mechanics:

1. The possible states of a system are those whose wavefunctions transform, under the operations of a group with respect to which the Hamiltonian is invariant, according to irreducible representations of this group. In other words, the eigenfunctions belonging to any one energy level form a basis for one of the representations of the group. For instance, in a cubic crystal there must be states $\Gamma_s$ whose wavefunction is left unchanged by all the operations of the cubic group. Each state is thus labelled by an irreducible representation, but at every **k**-point there are infinite bands belonging to the same irreducible representation (in the same way as in an atom there is an infinite number of *s*, *p*, *d* ... states), so that a band index based on the energy is still necessary, though in practice only one level of a given type is generally of interest. At a general point all the bands are of the same type: all levels are non-degenerate, excluding spin. Wavefunctions belonging to different irreducible representations are orthogonal, and all matrix elements of the energy operator between states belonging to different irreducible representations vanish.

2. There is one energy level for each irreducible representation and the degeneracy of each state is equal to the dimension of the corresponding irreducible representation. In the example of the full cubic group, the free electron level 48 times degenerate which we considered at $\mathbf{k} = 0$ splits, when a potential with the rotation and inversion symmetry of the crystal is introduced, into 20 different energy levels: one $\Gamma_s$ state, non degenerate; three $\Gamma_p$ states, each triply degenerate; two $\Gamma_d$ states, each doubly degenerate; three $\Gamma_{d'}$ states, each triply degenerate; one $\Gamma_f$ state, non degenerate; three $\Gamma_{f'}$ states, each triply degenerate; three $\Gamma_g$ states, each triply degenerate; two $\Gamma_h$ states, each doubly degenerate; one $\Gamma_i$ state, non degenerate; one $\Gamma_j$ state,

non degenerate. For the proper identification of the various levels, it is necessary to relate these crystal states to the free electron and to the atomic states, and we shall discuss this in the next Sections.

One may wonder why treat in detail the **k**-points of high symmetry, since they constitute an exceptional case. But energy extrema are normally found at these points, so that they are of extreme importance in semiconductors. Also, band structure calculations are simplified at these points.

## 6.2 Free electron correspondence

We want to find how the degenerate free electron states at point of high symmetry split under the influence of the crystal potential. This is better achieved by discussing a specific example and we shall consider again the $\Gamma$-states of a face-centred cubic lattice. These are (Section 5.1): a non-degenerate state of energy $E_0 = 0$; an eight times degenerate state of energy

$$E_1 = \frac{4\pi^2}{a^2}(1^2 + 1^2 + 1^2)$$

a six times degenerate state of energy

$$E_2 = \frac{4\pi^2}{a^2}(2^2 + 0^2 + 0^2)$$

etc. Each level will split into a number of crystal states belonging to irreducible representations of the $\Gamma$-group, but some degeneracy will still persist (it would be entirely removed only by consideration of magnetic effects). The problem is the same as for the general 48 times degenerate state treated in the last chapter, but now we shall consider it in detail.

In the free electron model the wavefunction of the state with energy $E_0$ is a constant. Each operation of the $\Gamma$-group leaves it unchanged and is thus represented by the one-dimensional matrix (1). The Character Table then shows that this state is of type $\Gamma_s$.

Since the vector $\mathbf{K}_1$ is $2\pi/a(\pm 1, \pm 1, \pm 1)$, the eight plane waves $e^{2\pi i/a(\pm x \pm y \pm z)}$ form a basis for a representation of the group, Take them in the order $(+++), (++-), (+-+), (-++), (+--), (-+-), (--+), (---)$. Each symmetry operation merely interchanges these functions and, as before, is represented by a square matrix whose elements are all zero except for one element in each row or column. For instance, the three operations of class $C_4{}^2$, which change $(x, y, z)$ into, respectively, $(x, -y, -z)$, $(-x, y, -z)$, $(-x, -y, z)$ are represented by

$$
\begin{vmatrix}
0&0&0&0&1&0&0&0\\
0&0&1&0&0&0&0&0\\
0&1&0&0&0&0&0&0\\
0&0&0&0&0&0&0&1\\
1&0&0&0&0&0&0&0\\
0&0&0&0&0&0&1&0\\
0&0&0&0&0&1&0&0\\
0&0&0&1&0&0&0&0
\end{vmatrix}
\quad
\begin{vmatrix}
0&0&0&0&0&1&0&0\\
0&0&0&1&0&0&0&0\\
0&0&0&0&0&0&0&1\\
0&1&0&0&0&0&0&0\\
0&0&0&0&0&0&1&0\\
1&0&0&0&0&0&0&0\\
0&0&0&0&1&0&0&0\\
0&0&1&0&0&0&0&0
\end{vmatrix}
\quad
\begin{vmatrix}
0&0&0&0&0&0&1&0\\
0&0&0&0&0&0&0&1\\
0&0&0&1&0&0&0&0\\
0&0&1&0&0&0&0&0\\
0&0&0&0&0&1&0&0\\
0&0&0&0&1&0&0&0\\
1&0&0&0&0&0&0&0\\
0&1&0&0&0&0&0&0
\end{vmatrix}
$$

The whole set of 48 $8 \times 8$ matrices is easily obtained and they form a representation of the $\Gamma$-group. This representation is reducible: we could have started with suitable linear combinations of the eight plane waves and we would have obtained the matrices in the form of a direct sum of irreducible representations $\Gamma_n$ associated with $\Gamma_n$-levels. To determine which $\Gamma_n$s occur in the direct sum, use is made of the theorem stating that the character of a reducible representation is, for each class, equal to the sum of the characters of the irreducible representations of which it is the direct sum. Since the character is the same for all operations in any one class, it is enough to write down one matrix for each class. The characters so obtained for the reducible representation are given in Table 6.2.1. One then picks up from the Character Table

**Table 6.2.1**

*Characters of reducible representation of Γ-group using* $e^{(\pm ix \pm iy \pm iz)}$ *as basis functions*

| *Class* | *Trace* | *Class* | *Trace* |
|---|---|---|---|
| $E$ | 8 | $J$ | 0 |
| $C_4^2$ | 0 | $JC_4^2$ | 0 |
| $C_4$ | 0 | $JC_4$ | 0 |
| $C_2$ | 0 | $JC_2$ | 4 |
| $C_3$ | 2 | $JC_3$ | 0 |

for the group (Table 6.1.1) those irreducible representations whose characters add up for each class to the trace given in Table 6.2.1. In this case, by inspection of the characters for classes $E$ and $JC_2$, it is immediately recognised that the irreducible representations are $\Gamma_s \oplus \Gamma_p \oplus \Gamma_{d'} \oplus \Gamma_f$. (For the zinc blende structure, from Table 6.1.3, the direct sum is $2\,\Gamma_s \oplus 2\,\Gamma_p$.) The corresponding basis functions are 1; $x, y, z$; $yz, zx, xy$; $xyz$. These are perhaps better seen as products of sine and cosines resulting from the combinations of the plane waves (omitting the factor $2\pi/a$): $\cos x \cos y \cos z$ (one state, $\Gamma_s$); $\cos x \cos y \sin z$, etc (three states, $\Gamma_p$); $\cos x \sin y \sin z$, etc (three states, $\Gamma_{d'}$); $\sin x \sin y \sin z$ (one state, $\Gamma_f$). Thus the original free electron eight times degenerate level splits into two non degenerate and two triply degenerate levels. Similarly, the six times degenerate level $E_2$ is found to split into $\Gamma_s \oplus \Gamma_p \oplus \Gamma_d$. It is immediately verified that, by using the symmetrised basis functions instead of the original plane waves, all matrices are transformed to a form in which they consist of blocks along the main diagonal. For example, for $E_1$, the operation changing $(x, y, z)$ into $(y, -z, x)$ is represented, using the plane waves as basis functions, by the matrix

$$\begin{vmatrix} 0&1&0&0&0&0&0&0 \\ 0&0&0&0&0&1&0&0 \\ 1&0&0&0&0&0&0&0 \\ 0&0&0&0&1&0&0&0 \\ 0&0&0&1&0&0&0&0 \\ 0&0&0&0&0&0&0&1 \\ 0&0&1&0&0&0&0&0 \\ 0&0&0&0&0&0&1&0 \end{vmatrix}$$

But when the symmetrised functions are used, the matrix becomes

$$\begin{vmatrix} 1&0&0&0&0&0&0&0 \\ 0&0&1&0&0&0&0&0 \\ 0&0&0&-1&0&0&0&0 \\ 0&1&0&0&0&0&0&0 \\ 0&0&0&0&0&-1&0&0 \\ 0&0&0&0&0&0&1&0 \\ 0&0&0&0&-1&0&0&0 \\ 0&0&0&0&0&0&0&-1 \end{vmatrix}$$

Also the problem of the determination of the basis functions, in particular when expressed as linear combinations of plane waves corresponding to each irreducible representation, can be tackled by similar methods. Consider an irreducible representation $r$, usually multi-dimensional. Let $i$ denote one particular row of each matrix. For instance, if it turns out that the basis functions are $x, y, z$, ($p$-like representation), then $i$ denotes $x$, or $y$, or $z$. Let the $ii$ diagonal element of the matrix representing the operation $R$ be $[D^{(r)}(R)]_{ii}$. Then the $i$ partner of the symmetrised combination is given by (see for instance Slater, 1965)

$$\Psi_i^{(r)} = \Sigma_R \, [D^{(r)}(R)]_{ii} R\Psi \tag{6.3}$$

where $\Psi$ is a plane wave (or orthogonalised plane wave, or augmented plane wave) $e^{i(\mathbf{k}+\mathbf{K}_n)\cdot\mathbf{r}}$ and the sum is over all the operations of the group. Although the combinations corresponding to different $i$ are generally different, they belong to the same eigenvalue and in most cases only one need be considered in energy determinations.

The symbol operating on $\Psi$ may be considered as a projection operator which projects out of any arbitrary function that portion which transforms according to the particular irreducible representation $r$ of the group in question. For the particular case in which $r$ is one-dimensional, $[D^{(r)}(R)]_{ii}$ reduces to the character for the class to which $R$ belongs and for the irreducible representation $r$. For instance, suppose we want the linear combination of the eight waves $e^{i(\pm x \pm y \pm z)}$ which transform according to the representation $\Gamma_{f}$. Take the wave $(+++)$. When we apply the operations of the cubic group, we find that it is brought into itself by the identity and by some operations of classes $C_3$ and $JC_2$; that it is brought into $(++-)$, or $(+-+)$, or $(-++)$ by the operations of classes $C_4$ and $JC_4^2$ and by some operations of classes $C_2$ and $JC_3$; that it is brought into $(+--)$, or $(-+-)$, or $(--+)$ by the operations of classes $C_4^2$ and $JC_4$ and by some operations of classes $C_3$ and $JC_2$; that it is brought into $(---)$ by the inversion and by some operations of classes $C_2$ and $JC_3$. By inspecting the Character Table 6.1.1 we find then from (6.3) that the required combination is

$$\begin{aligned}(+++) + (+--) + (-+-) + (--+) - (++-)\\ - (+-+) - (-++) - (---) = -8i \sin x \sin y \sin z\end{aligned}$$

Similarly it is found that one of the three combinations of the eight waves with symmetry $\Gamma_p$ is

$$\begin{aligned}(+++) + (+-+) + (++-) + (+--) - (-++)\\ - (--+) - (-+-) - (---) = 8i \sin x \cos y \cos z\end{aligned}$$

and one with symmetry $\Gamma_{d'}$ is

$$\begin{aligned}(+++) + (++-) + (--+) + (---) - (-++)\\ - (+-+) - (-+-) - (+--) = -8 \sin x \sin y \cos z\end{aligned}$$

The symmetrised combinations of plane waves at any **k**-point can also be found from the nearly free electron approximation. A practical example will illustrate the procedure. Consider the first free electron level at $W$ (Fig. 6.2) in aluminium. We have seen that, from the free electron band structure, illustrated in Fig. 5.2, there might conceivably be pockets of holes in the first band, or pockets of electrons in the third and fourth band, depending on whether the Fermi level for three electrons per unit cell falls above or below the energy at $W$. One may even think that, when the splitting due to the crystal field is considered, there might be pockets of both holes and electrons. Now in the nearly free electron approximation the four times degenerate level splits into three levels, two non-degenerate and one doubly degenerate. To see this, consider any particular $W$-point, for example, the point $W_1$ of co-ordinates (1, 2, 0) (omitting the factor $\pi/a$) and the three equivalent points discussed in connection with the $W$-group. The nearly free electron wave-functions at $W$ are thus linear combinations of the four plane waves with propagation vector equal to the four $W$-vectors:

$$\phi_W^{(\alpha)} = a\, e^{i(x+2y)} + b\, e^{i(x-2y)} + c\, e^{i(-x+2z)} + d\, e^{i(-x-2z)}$$

The coefficients are determined to first order by the equations ($E_0$ is the unperturbed free electron energy, $E_0 = 5\pi^2/a^2$)

$$\begin{aligned} E_0 a + V_2 b + V_1 c + V_1 d &= Ea \\ V_2 a + E_0 b + V_1 c + V_1 d &= Eb \\ V_1 a + V_1 b + E_0 c + V_2 d &= Ec \\ V_1 a + V_1 b + V_2 c + E_0 d &= Ed \end{aligned} \tag{6.4}$$

$V_1$, $V_2$ are the Fourier coefficients of the potential, or pseudopotential, corresponding to $\mathbf{K}_1$, $\mathbf{K}_2$. The roots of the compatibility equation are

$$\begin{aligned} E_1 &= E_0 - V_2 \text{ (twice)} \\ E_2 &= E_0 + V_2 + 2V_1 \\ E_3 &= E_0 + V_2 - 2V_1 \end{aligned}$$

Giving to $E$ the value $E_1$ in (6.4), two constants, say $a$, $c$, are arbitrary and we must have $b = -a$, $d = -c$, so that the two degenerate wavefunctions are

$$\begin{aligned} \Psi_1^{(1)} &= e^{i(x+2y)} - e^{i(x-2y)} \\ &= 4iy + \ldots \end{aligned}$$

and

$$\begin{aligned} \Psi_1^{(2)} &= e^{i(-x+2z)} - e^{i(-x-2z)} \\ &= 4iz + \ldots \end{aligned} \tag{6.5}$$

This doubly degenerate state with $p$ symmetry must be $W_p^{(2)}$; the two wave-functions (6.5) are the symmetrised combinations appropriate to this state, and $y$, $z$ may be taken as basis functions.

Giving to $E$ the value $E_2$, we find $a = b = c = d$, clearly a $W_s$ state. Giving to $E$ the value $E_3$, we find $a = b = -c = -d$, giving a wavefunction $4ix + \ldots$, which is $p$-type and non degenerate and so of symmetry $W_p^{(1)}$; $x$ may be taken as the simplest basis function.

One would expect $V_1$ and $V_2$ to be positive, and that $2V_1 \gg V_2$. Then $W_p^{(1)}$ ($E = E_3$) is the lowest level and most likely lies below the free electron value, $W_p^{(2)}$ ($E = E_1$) is the next higher, and $W_s$ ($E = E_2$) the highest, and probably considerably higher than the free electron level. It is reasonable to conclude that there are no holes in the first band because its top is very probably below the Fermi level, and that the $s$-band is too high to be occupied. For a fuller discussion see Harrison (1966); in the present connection the important point is that the perturbation calculation has provided the symmetrised combinations of plane waves for the states $W_s$, $W_p^{(1)}$, $W_p^{(2)}$. The calculation is perfectly similar when more plane waves are considered by adding to $\mathbf{k}$ any $\mathbf{K}_i$.

## 6.3 Atomic correspondence

When atoms come together in a lattice, the quantum numbers $l$, $m$ appropriate to atoms, cease to be good quantum numbers because the corresponding observables are no longer constants of the motion. The modulus and $z$-component of angular momentum are replaced as constants of motion by the components of crystal momentum. Instead of $s, p, d, \ldots$ levels we have, for each value of $\mathbf{k}$, states belonging to one of the irreducible representations of the point group of $\mathbf{k}$. Taking for instance $\mathbf{k} = 0$, each atomic level usually splits into a number of $\Gamma$-levels. To investigate this splitting (atomic correspondence) the same method is employed as for the free electron correspondence: reducible representations of the $\Gamma$-group are found using as basis functions the wavefunctions suitable for a free atom, namely the associated spherical harmonics $P_l^m(\cos\theta)\, e^{im\varphi}$, which are basis functions for the infinite rotation group. Then the reducible representations are expressed as direct sums of irreducible representations by making the character of each class in the reducible representations equal to the sum of its characters for the irreducible representations. In this way, wavefunctions in crystals are expanded in spherical harmonics, instead of plane waves. We have seen that both expansions are used in band calculations.

To find the characters of the various classes by using the spherical harmonics as basis functions, note that every operation of the cubic group, apart the inversion which simply brings in a factor $-1$ for harmonics of odd $l$, is a rotation of an angle $\Phi$ about an axis which, since there are no preferred directions for an atom, can always be taken to be the axis of the spherical harmonics. Consider then the $2l + 1$ functions belonging to a given $l$, for which $m = l, l - 1, \ldots, -l$. After the rotation the function $P_l^m(\cos\theta)\, e^{im\varphi}$ becomes $P_l^m(\cos\theta)\, e^{im(\varphi+\Phi)}$, and thus the $(2l + 1)$-dimensional representation provided by the basis function of order $l$ is diagonal and simply

$$\begin{pmatrix} e^{il\Phi} & 0 & \ldots & 0 \\ 0 & e^{i(l-1)\Phi} & \ldots & 0 \\ \cdot & \cdot & \ldots & \cdot \\ 0 & 0 & \ldots & e^{-il\Phi} \end{pmatrix}$$

with character

$$\sum_{m=-l}^{l} e^{im\Phi} = e^{-il\Phi}(1 + e^{i\Phi} + \ldots + e^{2il\Phi})$$
$$= \frac{\sin(l + \frac{1}{2})\Phi}{\sin \frac{1}{2}\Phi}$$

For classes $C_4^2$ and $C_2$, $\Phi = \pi$; for $C_4$, $\Phi = \frac{1}{2}\pi$; for $C_3$, $\Phi = \dfrac{2\pi}{3}$. Table 6.3.1 gives the resulting characters.

**Table 6.3.1**

*Characters of reducible representations of the Γ-group using the spherical harmonics as basis functions.*

| *Representation* | *Classes* | | | | | | | | | |
|---|---|---|---|---|---|---|---|---|---|---|
| | $E$ | $C_4^2$ | $C_4$ | $C_2$ | $C_3$ | $J$ | $JC_4^2$ | $JC_4$ | $JC_2$ | $JC_3$ |
| $l = 0$ | 1 | 1 | 1 | 1 | 1 | 1 | 1 | 1 | 1 | 1 |
| $l = 1$ | 3 | −1 | 1 | −1 | 0 | −3 | 1 | −1 | 1 | 0 |
| $l = 2$ | 5 | 1 | −1 | 1 | −1 | 5 | 1 | −1 | 1 | −1 |
| $l = 3$ | 7 | −1 | −1 | −1 | 1 | −7 | 1 | 1 | 1 | −1 |
| $l = 4$ | 9 | 1 | 1 | 1 | 0 | 9 | 1 | 1 | 1 | 0 |
| ... | . | . | . | . | . | . | . | . | . | . |

By comparing these characters with those in Table 6.1.1, it is immediately recognised that

a $s$-level becomes $\Gamma_s$;
a $p$-level becomes $\Gamma_p$;
a $d$-level splits into $\Gamma_d + \Gamma_{d'}$;
a $f$-level splits into $\Gamma_p + \Gamma_f + \Gamma_{f'}$;
a $g$-level splits into $\Gamma_s + \Gamma_d + \Gamma_{d'} + \Gamma_g$, etc.

It is seen that the suffixes given to the levels denote the atomic level of smallest $l$ to which each level is related in the atomic correspondence. If the wavefunction is expanded in spherical harmonics, the first $l$ that appears is the one corresponding to the suffix, and from the splitting scheme it is also apparent which orders of spherical harmonics are present in the develpoment. For instance, the expansion of a $\Gamma_s$ wavefunction contains one $s$, one $g$ function, etc. Exactly which combination (or combinations) of the $2l + 1$ spherical harmonics of order $l$ appear, can be determined only by a closer study of the symmetry. For the case of the cubic group, these combinations are called Kubic Harmonics (Von der Lage and Bethe 1947). We have seen that the spherical harmonic of lowest $l$ can be used with advantage as basis function for the corresponding irreducible representation.

For **k**-points of lower symmetry the splitting is different, but again the atomic correspondence is established immediately, given the Character Table of the point group of **k**. There are generally more crystal levels for any given $l$, and the wavefunctions of the various irreducible representations include more orders of spherical harmonics. At a **k**-point of no symmetry whatever, for which the point group reduces to the single identity operation, there is only one type of band, so that a $d$-level, for instance, splits into five bands, all of the same symmetry. Generally the wavefunctions at points of no symmetry include all the spherical harmonics. For instance, if we limit the expansion up to $l = 4$, there are $1 + 3 + 5 + 7 + 9 = 25$ terms. (Compare the case of $\Gamma_s$, for which there are only two terms up to $l = 4$.) However, even at points of no symmetry, often one, or a few coefficients are much larger than the others, so that the band can still loosely be called $s$-type, $p$-type, etc.; for instance, the '$d$-bands' of transition metals.

Within an $s$-band, or $p$-band, etc., the order of the levels, in particular the location of the band edges in the Brillouin zone, is found to be rather independent of the approximation employed in the calculation and of the assumed crystal potential. It is in fact determined primarily by the kinetic energy $k^2$ of the free electron approximation. For instance, the ordering of the levels of an $s$-band that we found for a f.c.c. structure in the tight binding approximation (Section 5.5.) is qualitatively the same whatever the method of calculation and whatever the potential, within reasonable limits. On the other hand, we have already seen that the relative position of levels of different types does sometimes depend on the crystal potential and is found to vary in a more or less systematic way from element to element in the same column of the Periodic Table and with the same crystal structure (for example in diamond, silicon, germanium, grey tin). The energy gap in such a series of semiconductors almost invariably decreases as the atomic number increases.

Another problem that can be solved just by comparing the relevant Character Tables is that of determining which irreducible representations along a certain direction in **k**-space, for example the $\Delta$-direction, are compatible with the irreducible representations at the end points, for example at $\Gamma$. For instance, from Tables 6.1.1 and 6.1.2 a band in a f.c.c. crystal which is of type $\Gamma_s$ at $k = 0$ must be of type $\Delta_s$ along the [1,0,0]-direction. A $\Gamma_p$ band gives rise to $\Delta_s + \Delta_p$; a $\Gamma_{d'}$ band to $\Delta_p + \Delta_{d'}$, etc.

Fig. 6.4 illustrates schematically the free electron and atomic correspondences for a face-centred cubic metal like Al at $\mathbf{k} = 0$. The crystal potential increases from left to right. The inner core levels $1s$, $2s$, $2p$ are not considered since, at the actual lattice spacing of Al, they remain practically sharp. It would perhaps be more appropriate to speak of a 'metallic', rather than of an atomic correspondence (Slater, 1934). Only if the lattice spacing were much reduced, would the lowest band correspond to the $1s$-level. From diagrams of this type, drawn for several **k**-points, valuable preliminary information is

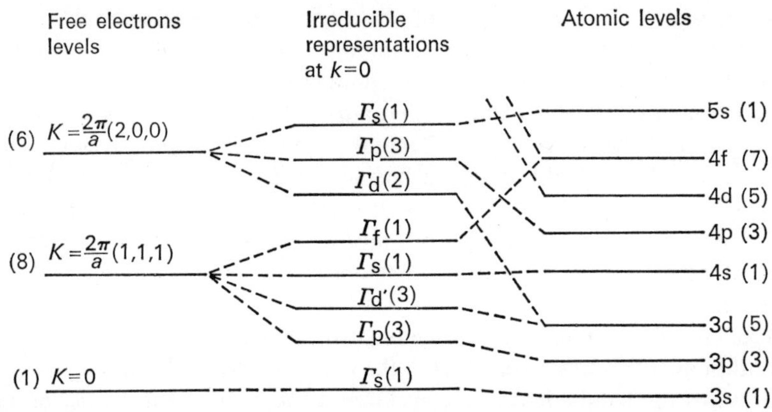


Fig. 6.4. Free electron and atomic correspondence at point $\Gamma$ for aluminium.

obtained on band structures. The next step then consists in carrying out at one end the simplest possible nearly free electron calculation, and at the other end the simplest tight binding one.

The discussion is more complicated for elements that have more than one electron outside closed shells, and for compounds. For instance, silicon and germanium have four electrons in states, respectively, $3s$, $3p$, and $4s$, $4p$. However, to obtain the valence 4, we need 4 electrons with spins in the same direction. For this, because of the Pauli principle, one $s$-electron must be promoted to the $p$-shell, obtaining one $s$, one $p_x$, one $p_y$, one $p_z$ electron. One then considers Bloch functions in the tight binding approximation formed from the $s$, $p_x$, $p_y$, $p_z$ orbitals of atoms on each of the two sub-lattices. There are eight states and, when the secular equation for the energy is solved, the solutions divide into two sets of four bands each. The lower set forms the valence band, holding the four outermost electrons for each Si or Ge atom. It is separated by an energy gap from the higher set which forms the conduction band, empty in a perfect crystal at 0 K. In the language of chemistry, the valence band is, roughly, formed from bonding combinations of the orbitals on the two sub-lattices, the conduction band from antibonding combinations. At $\mathbf{k} = 0$ there are, in silicon, in order of increasing energy:

1. A combination of *s*-orbitals symmetric about the centre of inversion midway between the atoms. This has $\Gamma_s$ symmetry and corresponds to the bottom of the valence band.
2. A similar symmetric combination of *p*-orbitals (symmetry $\Gamma_{d'}$), triply degenerate, corresponding to the top of the valence band.
3. An antisymmetric combination of *p*-orbitals (symmetry $\Gamma_p$), triply degenerate, corresponding to a secondary minimum of the conduction band.
4. An antisymmetric combination of *s*-orbitals (symmetry $\Gamma_f$), a higher level.

In a nearly free electron picture, the eight states derive from the eight times degenerate level of energy $K_1^2$ (Fig. 5.2). In germanium there is a similar state of affairs, except that the state $\Gamma_f$ is lower than $\Gamma_p$. $\Gamma_f$ is a sensitive level. One might have thought that $\Gamma_p$ should be below $\Gamma_{d'}$. That the opposite happens, is due to the particular symmetries of the orbitals combinations. The orbitals on adjacent atoms can combine to produce either a bonding (even) or an antibonding (odd) composite orbital. In the odd combination,

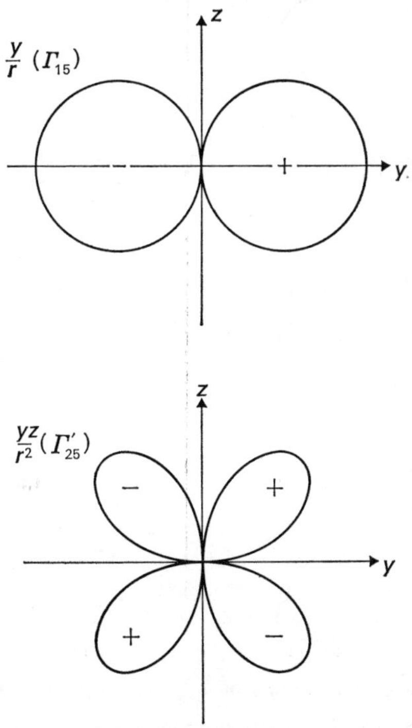


Fig. 6.5. Shape of orbitals forming basis functions for the irreducible representations $\Gamma_p(\Gamma_{15})$ and $\Gamma_{d'}(\Gamma'_{25})$.

a node of the wavefunction occurs at the midpoint between adjacent atoms. This node excludes electrons and thus prevents the formation of a covalent bond. For the formation of a bond, the combination must be even under inversion. This is true for $\Gamma_{d'}$, but not for $\Gamma_p$, so that the former state forms the top of the valence band.

Fig. 6.5 shows schematically the shape of the orbitals constituting the basis functions for $\Gamma_p$, $\Gamma_{d'}$. Even though the $\Gamma_{d'}$ orbitals are really *d*-functions (two azimuthal nodes), they can still form a basis for representing the *p*-like valence band edge.

As we have seen, the $\Gamma$-point group in the zincblende structure has only half as many elements as that in the diamond structure and it does not contain $\Gamma_{d'}$ as a three-dimensional representation, so that $\Gamma_p$ must form the

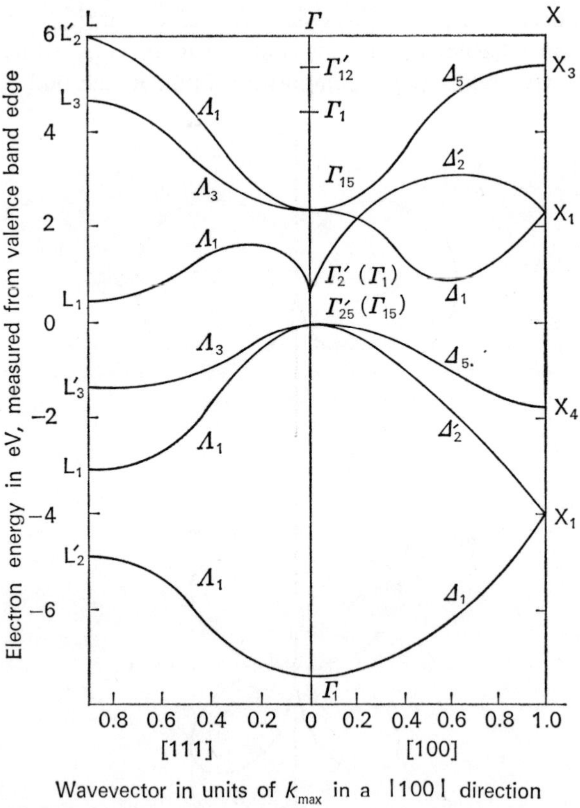


Fig. 6.6. Qualitative energy band structure of a crystal with the diamond lattice. Spin-orbit splitting is ignored. The symbols in parentheses represent the corresponding states for a crystal with the zincblende lattice. (After Long, 1968)

valence band edge. It is the lack of inversion symmetry that allows the $\Gamma_p$ combination to be bonding. Actually, $x$, $y$, $z$ and $yz$, $zx$, $xy$ are both valid sets of basis functions for $\Gamma_p$ in the zincblende structure, and the band edge consists of a mixture of them. In most of the semiconducting compounds with the zincblende structure, the lowest state in the conduction band has symmetry $\Gamma_s$.

Using the results of group theory and the kind of argument just illustrated for the $\Gamma$-group, with some help also from first-principle calculations and experimental results, it is possible to plot a qualitative band structure for a crystal with the diamond structure (Fig. 6.6—Long, 1968). The relative energies of the various levels that can be band edges are different in the different crystals. One may compare Fig. 6.6 with Fig. 5.2 and trace the derivation of the bands of the real crystals from the free electron bands (Herman, 1958).

The presence of the screw axis is important, because it causes the degeneracy of two bands at the point X. It is this degeneracy which produces the conduction band minima along the axes of fourfold symmetry. It is missing in crystals with the zincblende structure, which then have the conduction band edge at $k = 0$.

## 6.4 Time reversal and spin-orbit coupling

To complete the study of band structure that can be carried out by group theory, the effect of time reversal symmetry and of spin-orbit coupling must be considered. Time reversal means essentially reversing the direction of the motion of the electrons, that is changing $\mathbf{k}$ into $-\mathbf{k}$. The Hamiltonian does not depend on $t$, so that its eigenvalues cannot be different for the two possible directions of time flow. We have seen in Section 2.1 that it then follows from the fact that the Hamiltonian is real that, if $\Psi_{\mathbf{k}}(\mathbf{r})$ is an eigenfunction with eigenvalue $E(\mathbf{k})$, then $\Psi_{\mathbf{k}}^*(\mathbf{r})$ is also an eigenfunction belonging to the same eigenvalue and the equations (2.18) hold. When $\mathbf{k}$ is on the boundary of the Brillouin zone, time reversal symmetry may lead to additional degeneracies.

More important is the effect of spin-orbit coupling, though spin-orbit splitting can be ignored, particularly when only elements of low atomic number are involved, in a first attempt to determine the general features of a band structure. But it is often very important in determining the fine details at the band edges.

Spin-orbit coupling is only one of several relativistic effects, and, to deal with these, the Dirac equation must be used instead of the Schrödinger equation. Although the velocities of the electrons of interest in solid-state physics are always much less than the velocity of light, relativistic effects other than spin-orbit interaction can, as already mentioned (Section 5.7.4), be more important than one might expect. However, for the treatment of spin-orbit coupling the Pauli approximation is sufficient.

The Dirac theory, in the Pauli approximation, leads to the equation

$$H\Psi \equiv \left(\frac{p^2}{2m_0} + V + \frac{\hbar}{4m_0^2c^2}\,\boldsymbol{\sigma} \times \text{grad } V.\mathbf{p}\right)\Psi = E\Psi \tag{6.6}$$

where $\sigma_x$, $\sigma_y$, $\sigma_z$ are the Pauli matrices and the third term in the bracket represents the spin-orbit interaction. If

$$\Psi_{\mathbf{k}}(\mathbf{r}) = u_{\mathbf{k}}(\mathbf{r})\,e^{i\mathbf{k}.\mathbf{r}},$$

where $\Psi$ and $u$ are two-component spinors, the equation satisfied by $u$ is

$$\left[\frac{1}{2m_0}|\mathbf{p} + \hbar\mathbf{k}|^2 + V + \frac{\hbar}{4m_0^2c^2}\,\boldsymbol{\sigma} \times \text{grad } V.(\mathbf{p} + \hbar\mathbf{k})\right]u_{\mathbf{k}}(\mathbf{r}) = Eu_{\mathbf{k}}(\mathbf{r}) \tag{6.7}$$

This equation is an extension of (2.16) and can be treated by an extension of the $\mathbf{k}.\mathbf{p}$ approximation. The term

$$\frac{\hbar}{m_0}\mathbf{k}.\left[\mathbf{p} + \frac{\hbar}{4m_0c^2}\boldsymbol{\sigma} \times \text{grad } V\right] \tag{6.8}$$

is treated as a perturbation, assuming that we are dealing with a band edge at $\mathbf{k} = 0$. The other term

$$\frac{\hbar}{4m_0^2c^2}\boldsymbol{\sigma} \times \text{grad } V.\mathbf{p} \tag{6.9}$$

is important also at $\mathbf{k} = 0$. This term is large near nuclei, where $V$ varies rapidly. There the potential is almost spherically symmetric, so the treatment is the same as for isolated atoms (see, for instance, Slater, 1965), and (6.9) becomes

$$\frac{\hbar}{4m_0^2c^2}\,\frac{1}{r}\frac{dV}{dr}\mathbf{L}.\boldsymbol{\sigma} \tag{6.10}$$

where $\mathbf{L}$ is the orbital angular momentum operator $\mathbf{r} \times \mathbf{p}$. Thus the spin-orbit splitting energy in an elemental solid is not very different from that in the free atom.

We have seen that, because of time inversion symmetry,

$$E(-\mathbf{k}) = E(\mathbf{k})$$

But time inversion reverses the spin direction, as can be seen by considering spin as involving a circulating electric current, so that, when spin is considered,

$$E(-\mathbf{k})\uparrow = E(\mathbf{k})\downarrow$$

On the other hand, the inversion operator has no effect on spin. If the crystal then has inversion symmetry,

$$E(-\mathbf{k})\uparrow = E(\mathbf{k})\uparrow$$

and thus

$$E(\mathbf{k})\uparrow = E(\mathbf{k})\downarrow:$$

for these crystals the double degeneracy due to spin is present at all **k**-points. However, if the structure has no centre of inversion (zincblende structure), the degeneracy is removed, but still

$$E(-\mathbf{k}) = E(\mathbf{k})$$

Fig. 6.7 indicates schematically the top of one of the valence bands in Ge and InSb respectively and illustrates the point in question. It is found that in InSb and InAs there are eight maxima in the [1,1,1]-directions. Their distance from $\mathbf{k} = 0$ is very small, about 1 % of the size of the Brillouin zone, and the energy is only about $10^{-3}$ eV higher than at $\mathbf{k} = 0$.

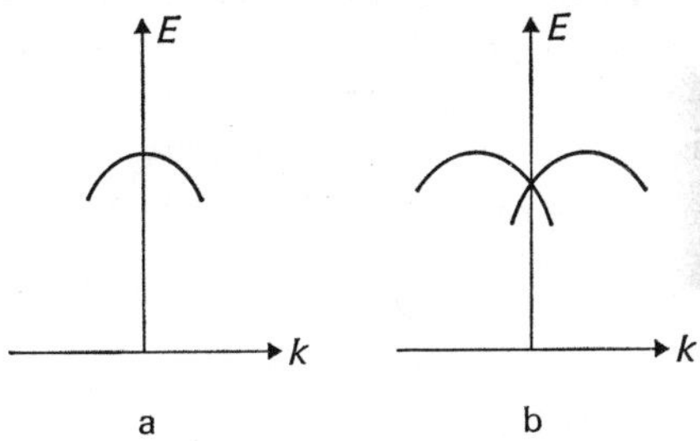


Fig. 6.7. Top of one of the three valence bands (a) in a crystal with the diamond structure (e.g. Ge), (b) in a crystal with the zincblende structure (e.g. InSb). All states in (a) are doubly degenerate due to spin. The lack of an inversion centre removes the degeneracy in (b).

From the point of view of group theory, consideration of spin has the effect of changing every **k**-point group into a 'double group'. To understand how this happens, we recall that the two possible orientations of spin are conveniently denoted by the Pauli functions $\alpha$ and $\beta$ and by the quantum numbers $m_s = \pm\frac{1}{2}$. By analogy with orbital momentum, one may roughly say that $\alpha$ behaves like a function $e^{\frac{1}{2}i\phi}$, where $\phi$ is an angle of rotation, and $\beta$ like $e^{-\frac{1}{2}i\phi}$.

The complication arises that, if $\phi$ increases by $2\pi$, both $\alpha$ and $\beta$ are multiplied by $e^{\pi i} = -1$, that is, they change sign. Thus, for every rotation in the range 0 to $2\pi$, we must have another rotation of an angle between $2\pi$ and $4\pi$, whose effect on any function of coordinates is the same as that of the original rotation, but whose effect on a spin function is to change its sign. Therefore, in general, if the operations of a point group operate on a spin function, for each operation involving a rotation, there will be another one, corresponding to a rotation greater (or smaller) by $2\pi$, which gives a function of opposite sign. There are then twice as many operations to consider as with the same group when spin is not considered: this is the reason for the name 'double group'.

For each double group one must investigate the classes and find the irreducible representations and the basis functions. Of course, the irreducible representations and basis functions found for the single group hold also for the double group, but there are other irreducible representations, called the extra representations, whose basis functions involve the spin functions. These extra representations are those to be used for the labelling of the band when spin is considered: the wavefunctions must be basis functions for one of the extra irreducible representations of the double group. For instance, the double $\Gamma$-group has six additional classes (and irreducible representations) besides the ten classes of the single group. The new irreducible representations are denoted by $\Gamma_6^+$, $\Gamma_6^-$, $\Gamma_7^+$, $\Gamma_7^-$ (all two-dimensional), $\Gamma_8^+$, $\Gamma_8^-$ (both four-dimensional). The superscript indicates whether the corresponding basis functions are even or odd with respect to the inversion. The distinction between + and − is then missing in zincblende type crystals. It is found that the basis functions for $\Gamma_6^+$ are just the two spin functions $\alpha$, $\beta$; for $\Gamma_6^-$ they are

$$z\alpha + (x + iy)\beta$$

and

$$-z\beta + (x - iy)\alpha$$

for $\Gamma_8^-$ they are

$$(x + iy)\alpha$$

$$z\alpha - \frac{x + iy}{2}\beta$$

$$z\beta + \frac{x - iy}{2}\alpha$$

$$(x - iy)\beta, \text{ etc.}$$

A band which, without considering spin, was of type $\Gamma_s$ becomes, when spin is considered, of type $\Gamma_6^+$; $\Gamma_p$ becomes $\Gamma_6^- + \Gamma_8^-$, etc. The levels which, without spin, are threefold degenerate at $\Gamma$ double in number when spin is considered and divide into a twofold degenerate level ($p_{\frac{1}{2}}$ in spectroscopic

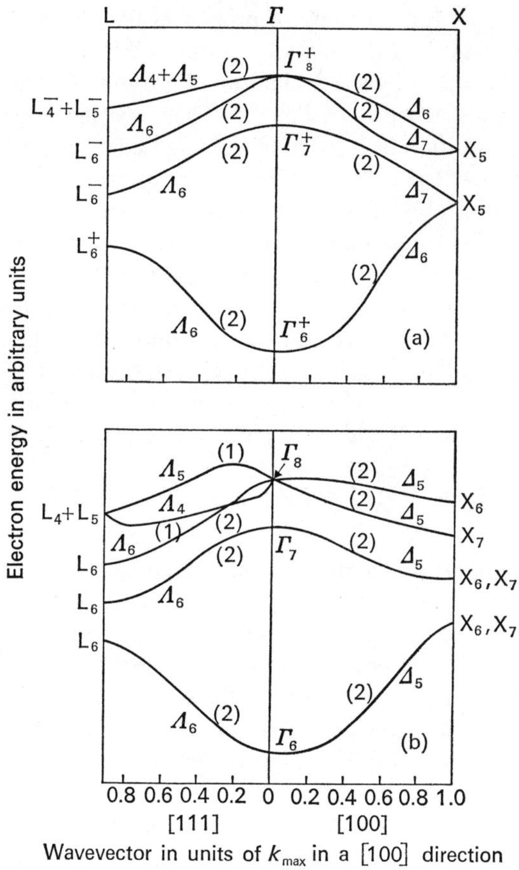


Fig. 6.8. Qualitative valence band structure of (a) diamond and (b) zincblende type crystals including spin-orbit interaction. The degeneracies of the $E(k)$ curves are indicated by the numbers in parenthesis. (After Long)

notation) and a fourfold degenerate level ($p_{\frac{3}{2}}$). Fig. 6.8 (also from Long, 1968) shows schematically the valence bands in crystals with the diamond structure when the effect of spin has been taken into consideration. For further details see Koster (1957) or Slater (1965).

## 6.5 Structure of band edges in semiconductors

In semiconductors, one has generally to do only with electrons or holes within a narrow range of energies above the energy at the band edge. Thus, only the energy surfaces near the **k**-points where the energy is an extremum are of interest. Although a calculation of the entire band structure is desirable, and in fact necessary for the study of properties like the optical absorption, the possible forms of the $E(\mathbf{k})$ relations in the vicinity of the edges can often be determined accurately by a combination of group theory and perturbation theory, as in the construction of Fig. 6.8. Then, specific experiments permit to determine the position in **k**-space of the band edges and to decide among the possible forms of the energy surfaces given by perturbation theory. This consists almost invariably of the generalised $\mathbf{k}.\mathbf{p}$ approximation, the perturbing terms being (6.8) and (6.9), and the dispersion relation is expressed through the components of the inverse mass tensor at the edge.

Semiconductors show a great variety of band structures, depending on the position of the edges in **k**-space and on the symmetry of the bands at these edges. We have seen that, apart from spin effects, the edge must be at a point of symmetry (Section 6.1). The simplest case is that of a non-degenerate $s$-band with edge at $\mathbf{k} = 0$ and isotropic effective mass. Then

$$E(\mathbf{k}) = E_0 + \frac{h^2}{2m^*}(k_x^2 + k_y^2 + k_z^2) \tag{6.11}$$

The constant energy surfaces are spheres. The conduction band of the III–V semiconducting compounds of the type of GaAs is of this type, and also that of some II–VI compounds with the zincblende structure. The region of **k**-space within which the energy may be considered both spherically symmetric and increasing quadratically with **k** is however, generally small. Particularly in InSb the deviation from a parabolic law is very marked.

Another possibility is for the edge to be at a symmetry point on the boundary of the Brillouin zone, for instance at the point $L$ (Fig. 5.1) for f.c.c. structures. In this case the symmetry of the energy surfaces will generally be cylindrical about an axis, taken as the $x$-axis, perpendicular to the hexagonal face of the Brillouin zone at $L$. Let the coordinates of $L$ be (0,0,0) and take the energy at the edge as zero. Then

$$E = \tfrac{1}{2}\hbar^2\left(\frac{k_x^2}{m_l} + \frac{k_y^2 + k_z^2}{m_t}\right) \tag{6.12}$$

$m_l$ is the longitudinal effective mass, $m_t$ the transverse effective mass. We have seen in Section 4.6 how these masses can be determined by means of cyclotron resonance experiments. The constant energy surfaces are illustrated in Fig. 6.9. The dotted part represents the continuation of the surface for values of **k** outside the Brillouin zone, and is equivalent to the full half-surface centred at the opposite face of the Brillouin zone. Each of the eight hexagonal faces has an energy minimum at its centre, but, since the two

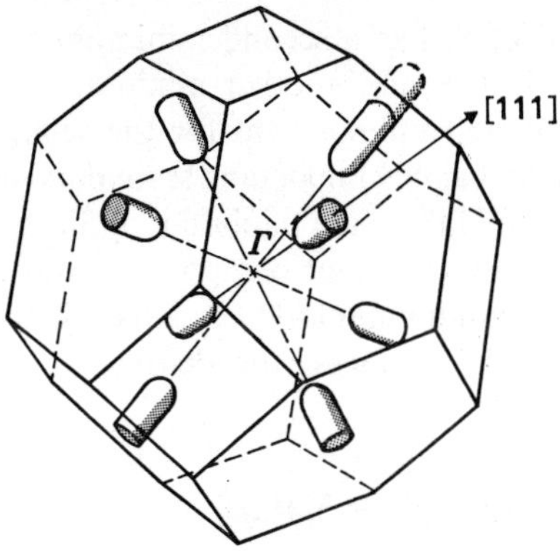


Fig. 6.9. Constant energy surfaces near the conduction band edge of germanium.

points on opposite faces are equivalent, the number of minima is 4, not 8. These minima are called 'valleys'. The conduction band of Ge has a structure of this type, with

$$m_l = 1{\cdot}59\, m_0,$$

$$m_t = 0{\cdot}082\, m_0$$

This band has also a second minimum at $\mathbf{k} = 0$, of the type previously considered, with energy 0·15 eV higher than at $L$. At this second minimum the practically isotropic $m^*$ is $0{\cdot}04\, m_0$. As indicated in Fig. 6.8 there may be a third set of minima along the $\Delta$-directions. The semiconducting compounds PbS, PbSe, PbTe, with the NaCl structure, have both the valence band and the conduction band edge similar to the conduction band edge of Ge.

Or the edge can be along a symmetry direction not on the zone surface, for instance along the $\Delta$-direction for a cubic structure. There are then six equivalent minima centred at $(k_0, 0, 0)$, $(-k_0, 0, 0)$, $(0, k_0, 0)$, $(0, -k_0, 0)$, $(0, 0, k_0)$, $(0, 0, -k_0)$. Again there will be generally a longitudinal and a transverse effective mass, and the equation of the constant energy surfaces centred at $(k_0, 0, 0)$ will be

$$E = E_0 + \tfrac{1}{2}\hbar^2\left[\frac{(k_x - k_0)^2}{m_l} + \frac{k_y^2 + k_z^2}{m_t}\right] \tag{6.13}$$

with similar expressions for the other five valleys. The conduction band of Si has this kind of structure with

$$m_l = 0{\cdot}90\, m_0$$
$$m_t = 0{\cdot}19\, m_0$$

The conduction band of Si has a second minimum at $L$, corresponding to the lowest minimum in Ge, and another minimum at $\mathbf{k} = 0$, but they are much higher in energy and unimportant for the semiconducting properties of the substance. GaAs has an important secondary minimum of the conduction band of the type of the main minimum of Si, about 0·36 eV above the lowest minimum at $k = 0$. In other compounds with the zincblende structure, such as AlSb, the minimum along the Δ-axes is the lowest.

For ellipsoidical energy surfaces, the density of states has, instead of (2.54), the form

$$dn_s = \frac{2^{\frac{1}{2}}}{\pi^2\hbar^3}(m_l m_t^2)^{\frac{1}{2}}(E - E_0)^{\frac{1}{2}}\, dE \tag{6.14}$$

The quantity $(m_l m_t^2)^{\frac{1}{3}}$ is called the density of states effective mass.

The most important semiconductors, such as Si, Ge and several III–V compounds, have a valence band deriving from a $p$ atomic level, with its triply degenerate edge at $\mathbf{k} = 0$. We now consider more in detail the valence band edge of Ge, taking also spin into account. The valence band of silicon is similar, and, apart from the small differences illustrated in Fig. 6.7, so is that of most compounds with the zincblende structure (InSb, InAs, GaSb, GaAs, AlSb, InP, GaP, CdTe, ZnTe, ZnSe).

We have already discussed the fact that in crystals with the diamond structure the edge has symmetry $\Gamma_{d'}$ despite the fact that the electrons are prevalently $p$-type. Taking spin into account, but in the absence of spin-orbit splitting, the hole level at $\mathbf{k} = 0$ is six times degenerate: the spin angular momentum may be parallel to the orbital momentum ($j = \frac{3}{2}$ – four levels), or antiparallel to it ($j = \frac{1}{2}$ – two levels). Spin-orbit interaction partially removes the degeneracy by lowering the two bands with antiparallel spin with respect to the four bands with parallel spin.

We have seen that, to study the form of the energy surfaces near a band edge, the $\mathbf{k.p}$ approximation is appropriate. Consider then a complete orthonormal set of periodic wavefunctions at $\mathbf{k} = 0$, $|u_{Ji}\rangle$, where $J$ indicates different energies and $i$ refers to the degenerate functions for a given energy. First, neglecting spin, the index $i$ takes three values for the three states of the triply degenerate $\Gamma_{d'}$-level. The difference between the energy at $\mathbf{k}$ and the energy at 0 is then determined (see (4.48)), for small $\mathbf{k}$, by the energy matrix

$$\langle u_{Ji}|H_{\mathbf{k}.\mathbf{p}}|u_{Jj}\rangle = \frac{\hbar^2}{m_0^2}\Sigma'_{L,n}\frac{\langle u_{Ji}|\mathbf{k}.\mathbf{p}|u_{Ln}\rangle\langle u_{Ln}|\mathbf{k}.\mathbf{p}|u_{Jj}\rangle}{E_J - E_L} \tag{6.15}$$

$i, j, n$ represent basis functions, which we can take as $x, y, z$; the prime on the summation means that $L = J$ is excluded.

Consideration of the symmetry of the problem shows that the energy secular equation must have the form

$$\begin{vmatrix} Ak_x^2 + B(k_y^2 + k_z^2) - E & Ck_xk_y & Ck_xk_z \\ Ck_yk_x & Ak_y^2 + B(k_z^2 + k_x^2) - E & Ck_yk_z \\ Ck_zk_x & Ck_zk_y & Ak_z^2 + B(k_x^2 + k_y^2) - E \end{vmatrix} = 0 \tag{6.16}$$

where the parameters $A$, $B$, $C$ are sums over matrix elements of the type of that in (6.15). Along directions of no particular symmetry there are three non-degenerate bands. Along the [1,0,0]-direction there is a twofold degenerate band with

$$E = Bk_x^2$$

and a non-degenerate band with

$$E = Ak_x^2.$$

Along the [1,1,1]-direction there is a two-fold degenerate band with

$$E = (A + 2B - C)k_x^2$$

and a non-degenerate band with

$$E = (A + 2B + 2C)k_x^2$$

To visualise the shape of the constant energy surfaces, consider their sections in the $k_z = 0$ plane. These curves consist of circles $E = B(k_x^2 + k_y^2)$ and of curves satisfying

$$E = \tfrac{1}{2}(A + B)(x^2 + y^2) \pm [\tfrac{1}{4}(A - B)^2(x^2 - y^2)^2 + C^2x^2y^2]^{\frac{1}{2}}$$

where $x$, $y$ have been written for $k_x$, $k_y$. They have been plotted in Fig. 6.10 ('warped' energy surfaces). All three curves correspond to the same value of the energy. The curve farthest from $\mathbf{k} = 0$ corresponds to the largest effective mass.

We now introduce the spin terms in the $\mathbf{k}.\mathbf{p}$ scheme. In fact the term proportional to $\mathbf{k}.\boldsymbol{\sigma} \times \text{grad}\, V$ is negligible near $\mathbf{k} = 0$ compared to the term proportional to $\mathbf{p}.\boldsymbol{\sigma} \times \text{grad}\, V$. Consideration of spin doubles each of the three states previously considered, so that we have a $6 \times 6$ energy matrix

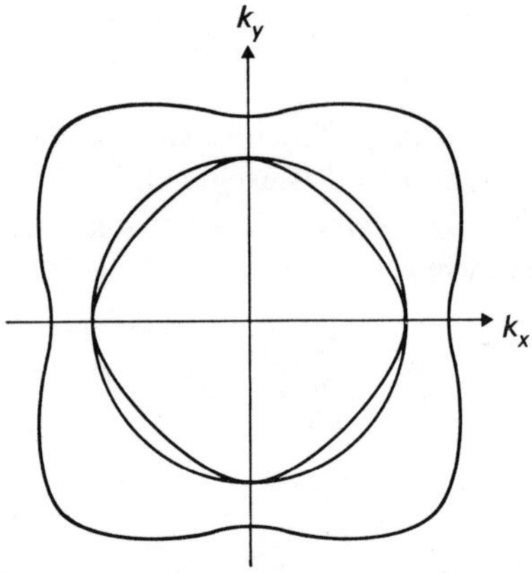


Fig. 6.10. Warped constant energy surfaces in the valence band of germanium (schematic).

with rows and columns labelled (with obvious symbols) by $x^+$, $y^+$, $z^+$, $x^-$, $y^-$, $z^-$, of the form

$$\begin{vmatrix} H_{\mathbf{k}\cdot\mathbf{p}} & 0 \\ 0 & H_{\mathbf{k}\cdot\mathbf{p}} \end{vmatrix}$$

To this we must add the matrix of the spin orbit interaction. This matrix is diagonal in the $j$, $m_j$ representation where it has the form

$$\begin{vmatrix} -2D/3 & & & & & \\ & -2D/3 & & & & \\ & & D/3 & & & \\ & & & D/3 & & \\ & & & & D/3 & \\ & & & & & D/3 \end{vmatrix}$$

where $D$ is the energy difference between the two levels $j = 3/2$ and $j = \frac{1}{2}$.

In our representation the spin-orbit matrix has instead the form (Kane, 1956)

$$H_{so} = -\frac{D}{3}\begin{vmatrix} 0 & i & 0 & 0 & 0 & -1 \\ -i & 0 & 0 & 0 & 0 & i \\ 0 & 0 & 0 & 1 & -i & 0 \\ 0 & 0 & 1 & 0 & -i & 0 \\ 0 & 0 & i & i & 0 & 0 \\ -1 & -i & 0 & 0 & 0 & 0 \end{vmatrix}$$

$H_{k.p}$ and $H_{so}$ must be added, and the equation

$$(H_{k.p} + H_{so})\Psi = E\Psi$$

solved. $D$ is small in Ge (0·30 eV) and even smaller in Si (0·04 eV), thus the shape of the bands previously deduced gets modified only near $\mathbf{k} = 0$.

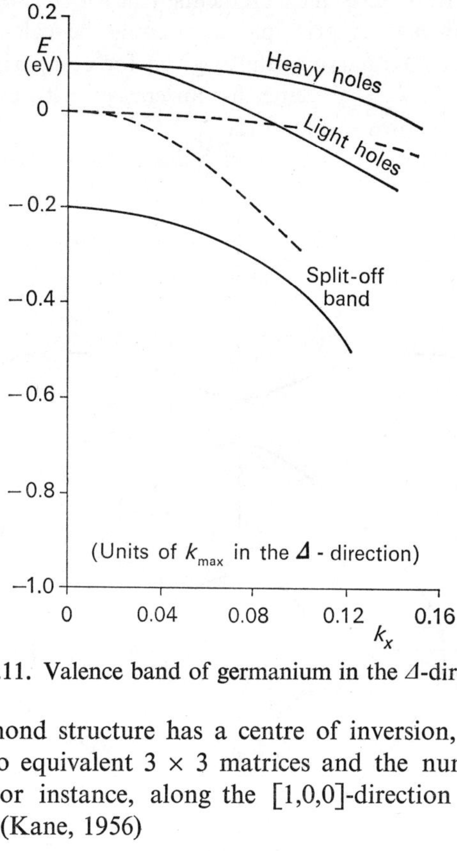


Fig. 6.11. Valence band of germanium in the $\Delta$-direction.

Since the diamond structure has a centre of inversion, the matrix must factorise into two equivalent 3 × 3 matrices and the number of bands is not increased. For instance, along the [1,0,0]-direction (writing again $x$ for $k_x$) one finds (Kane, 1956)

$$\begin{vmatrix} Bx^2 - E & D/3 & iD/3 \\ D/3 & Ax^2 - E & -iD/3 \\ -iD/3 & iD/3 & Bx^2 - E \end{vmatrix} = 0$$

At $\mathbf{k} = 0$ the roots are $E = \frac{D}{3}$ (twice) and $E = -\frac{2D}{3}$ as for an atom. Along the $\Delta$-axis one root is $E = \frac{D}{3} + Bx^2$ and the other two energies are roots of the equation

$$E^2 - \left(Ax^2 + Bx^2 - \frac{D}{3}\right)E + ABx^4 - \frac{AD}{3}x^2 - \frac{2D^2}{9} = 0$$

These bands are illustrated in Fig. 6.11, the parameters having been taken from cyclotron resonance measurements (they determine of course the effective mass tensor); in principle they could be calculated. The bands without spin-orbit interaction are also given for comparison. The constant energy curves in the $k_z = 0$ plane, for an energy just below the top of the split-off band, are shown in Fig. 6.12.

We see that there are two types of holes in Si and Ge (heavy and light

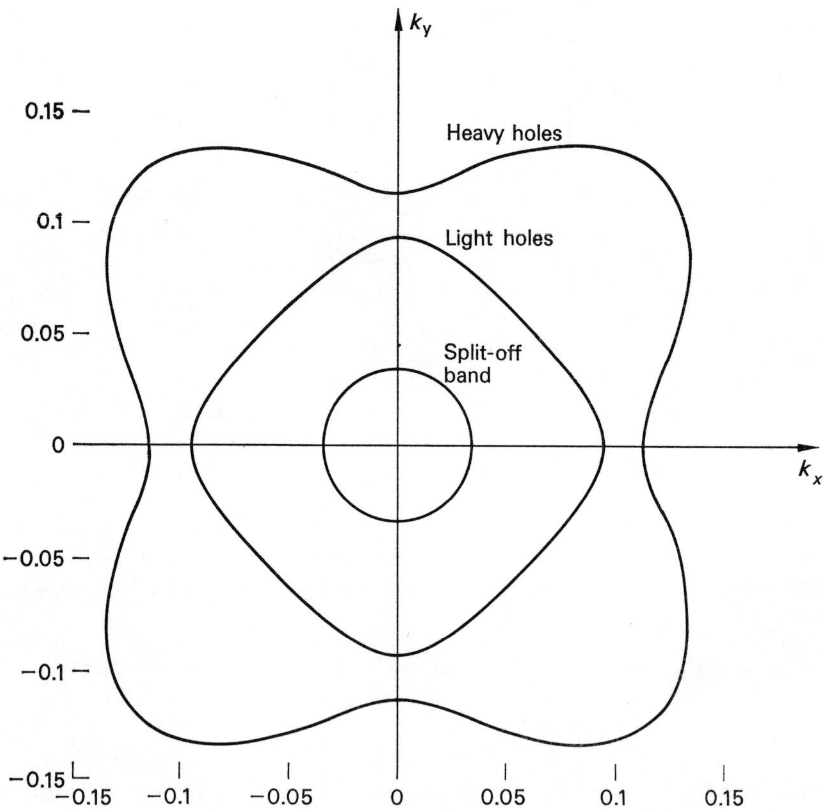


Fig. 6.12. Constant energy curves in the plane $k_z = 0$. Valence band of silicon with spin-orbit coupling.

$E = 0{\cdot}045$ Rydberg. $k$ in (Bohr radii)$^{-1}$.

holes), with different effective masses. Transitions between the two bands occur very readily, thus the light holes have a short lifetime, so that they are not observed in drift mobility experiments. The lifetime however is sufficiently long to detect light holes in cyclotron resonance experiments. The third band is called the split-off band. In Ge it is about 0·3 eV below the other two. The average mass of the heavy holes is 0·28 $m_0$ in Ge and 0·49 $m_0$ in Si; for the light holes the values are 0·044 in Ge and 0·16 in Si; for the split-off band, 0·077 in Ge, 0·245 in Si. The density of states effective mass is 0·39 in Ge and 0·55 in Si. The split-off band is spherically symmetric, but the light and heavy holes energy surfaces are warped. They must of course have full cubic symmetry.

In grey tin the ordering of the states at $\mathbf{k} = 0$ is different: the $\Gamma_f$ state, which in Ge belongs to the conduction band, lies below $\Gamma_{d'}$ which in Ge is the top of the valence band. Referring to the $\mathbf{k}.\mathbf{p}$ approximation (Section 4.5), the reversal of sign of the energy difference causes both bands to reverse their curvature, so that the light holes valence band of Ge becomes in grey tin a conduction band, but is still degenerate at $\mathbf{k} = 0$ with the heavy holes band. Grey tin is thus a semi-metal, with an energy gap exactly equal to zero. The conduction band of Ge at $\mathbf{k} = 0$ becomes another split off valence band. The minimum of the conduction band at L still exists in grey tin, but it lies about 0·09 eV above the minimum at $\mathbf{k} = 0$. The compound HgTe, with the zincblende structure, has a similar band structure.

# References

Further references to original papers can be found in the textbooks and review articles listed below.


ALDER, B. (Editor) (1968). *Energy Bands of Solids*. Vol. 8 of *Methods in Computational Physics*. Academic Press, New York.

ANDERSON, P. W. (1963). *Concepts in Solids*. Benjamin, New York.

BLOCH, F. (1928). *Zeit. Phys.*, **52**, 555.

BLOUNT, E. I. (1962). *Solid-State Physics*. Vol. 13, p. 306. Academic Press, New York.

BORN, M. and HUANG, K. (1954). *Dynamical Theory of Crystal Lattices*. Clarendon Press, Oxford.

BRILLOUIN, L. (1946). *Wave Propagation in Periodic Structures*. McGraw-Hill, New York.

CALLAWAY, J. (1964). *Energy Band Theory*. Academic Press, New York.

COHEN, M. L., and HEINE, V. (1970). *Solid State Physics,* Vol. 24, p. 38. Academic Press, New York.

CORNWELL, J. F. (1969). *Group Theory and Electronic Energy Bands in Solids*. North-Holland.

EDWARDS, S. F. (1967). *Advances in Physics*, **16**, 359.

EWALD, P. P. (1921). *Ann. Physik*, **64**, 253.

HAM, F. S. (1955). *Solid-State Physics,* Vol. 1, p. 127. Academic Press, New York.

HARRISON, W. A. (1966). *The Application of Pseudopotentials to Metals*. Benjamin, New York.

HEINE, V. (1960). *Group Theory in Quantum Mechanics*. Pergamon Press, Oxford.

HEINE, V. (1969). Article on *The Electronic Structure of Metals* in *The Physics of Metals,* Vol. 1; edited by J. M. Ziman. Cambridge University Press.

HEINE, V. (1970). *Solid State Physics,* Vol. 24, p. 1. Academic Press, New York.

HERMAN, F. (1958). *Rev. Mod. Phys.*, **30**, 102.

HERMAN, F. KORTUM, R. L., KUGLIN, C. D. and SHORT, R. A. (1967). In *Quantum Theory of Atoms, Molecules and Soild State*. Academic Press, New York.

HERRING, C. (1940). *Phys. Rev.*, **57**, 1169.

HOWARTH, D. J. (1955). *Phys. Rev.*, **99**, 469.

JONES, H. (1960). *The Theory of Brillouin Zones and Electronic States in Crystals.* North-Holland, Amsterdam.

KANE, E. O. (1956). *J. Phys. Chem. Solids*, **1**, 82.
KITTEL, C. (1954). *Amer. J. Phys.*, **22**, 250.
KITTEL, C. (1963). *Quantum Theory of Solids*. Wiley, New York.
KNOX R. S. (1963). *Theory of Excitons*. Academic Press, New York.
KOHN, W. (1957). *Phys. Rev.*, **105**, 509; **110**, 857.
KOHN, W. (1959). *Phys. Rev.*, **115**, 809.
KOHN, W. and ROSTOKER, N. (1954). *Phys. Rev.*, **94**, 1111.
KORRINGA, J. (1947). *Physica*, **13**, 392.
KOSTER, G. F. (1957). *Solid-State Physics*, Vol. 5, p. 174. Academic Press, New York.
LANDAU, L. D. and LIFSCHITZ, E. M. (1958). *Quantum Mechanics*, p. 471. Pergamon Press, Oxford.
LIFSCHITZ, E. M., AZBEL', M. YA., and SLUTSKIN, A. A. (1961). *Zh. eksp. teor. Fiz.*, **43**, 1464.
LINDHARDF, J. (1954). *Kgl. Danske Videnskab. Selskab. Mat-Rys. Medd*, **28**, 8.
LIPSON, H. and COCHRAN, W. (1953). *The Determination of Crystal Structures*. G. Bell and Sons, London.
LONG, D. (1968). *Energy Bands in Semiconductors*. Interscience, New York.
LOUCKS, T. (1967). *Augmented Plane Wave Method*. Benjamin, New York.
LÖWDIN, Per-O. (1950). *J. Chem. Phys.*, **18**, 365.
LUTTINGER, J. M. and KOHN, W. (1955). *Phys. Rev.*, **97**, 869.
MOTT, N. F. (1961). *Phil. Mag.*, **6**, 287.
MOTT, N. F. (1967). *Advances in Physics*, **16**, 49.
MOTT, N. F. and JONES, H. (1936). *The Theory of the Properties of Metals and Alloys*. Dover, New York.
NUSSBAUM, A. (1966). *Solid State Physics*, Vol. 18, p. 165. Academic Press, New York.
PEKAR, S. I. (1963). *Research in Electron Theory of Crystals* (translated from the Russian). U.S. Atomic Energy Commission, Washington, DC.
PENDRY, J. B. (1968). *Journal of Physics*, C, **1**, 1065.
PHARISEAU, P. and ZIMAN, J. M. (1963). *Phil. Mag.*, **8**, 1487.
PHILLIPS, F. C. (1963). *An Introduction to Crystallography*. Wiley, New York.
PHILLIPS, J. C. and KLEINMAN, L. (1959). *Phys. Rev.*, **116**, 287.
PINCHERLE, L. (1960). *Reports of Progress in Physics*, Vol. 23, p. 355.
PINES, D. (1963). *Elementary Excitations in Solids*. Benjamin, New York.
RAIMES, S. (1961). *The Wave Mechanics of Electrons in Metals*. North-Holland, Amsterdam.
SCHIFF, L. I. (1949). *Quantum Mechanics*, p. 153. McGraw-Hill, New York.
SEITZ, F. (1940). *Modern Theory of Solids*. McGraw-Hill, New York.
SLATER, J. C. (1934). *Rev. Mod. Phys.*, **6**, 209.
SLATER, J. C. (1951). *Phys. Rev.*, **81**, 385.
SLATER, J. C. (1957). Article in Vol. 19 of *Handbuch der Physik*. Springer, Berlin.
SLATER, J. C. (1965, 1967). *Quantum Theory of Molecules and Solids*, Vols. 2 and 3. McGraw-Hill, New York.
SLATER, J. C., KOSTER, G. F. and WOOD, J. H. (1962). *Quart. Rep. Solid State and Mol. Theory Group*, Massachusetts Institute of Technology, No. 43.

SMITH, R. A. (1961). *Wave Mechanics of Crystalline Solids.* Chapman and Hall, London.

SOMMERFELD, A. and BETHE, H. (1933). Article in *Handbuch der Physik*, Vol. 26, II. Springer, Berlin.

VON DER LAGE, F. C. and BETHE, H. A. (1947). *Phys. Rev.*, **71**, 612.

WANNIER, G. H. (1937). *Phys. Rev.*, **52**, 191.

WANNIER, G. H. (1959). *Elements of Solid State Theory.* Cambridge University Press, London.

WEINREICH, G. (1965). *Solids: Elementary Theory for Advanced Students.* Wiley, New York.

WILSON, A. H. (1953). *The Theory of Metals.* Cambridge University Press, New York.

WIGNER, E. and SEITZ, F. (1933). *Phys. Rev.*, **43**, 804.

ZAK, J. (1970). *Comments on Solid State Physics,* **2**, 209; **3**, 28.

ZIMAN, J. M. (1964). *Principles of the Theory of Solids.* Cambridge University Press, London.

ZIMAN, J. M. (1965). *Proc. Phys. Soc,* **86**, 337.

# Index